The Restoration of the Tidal Thames

Thameswater, the Thames Water Authority's new pollution control launch seen here during trials in the Thames. Photograph courtesy of the Thames Water Authority.

The Restoration of the Tidal Thames

Leslie B Wood

Consultant Editor: Professor D V Ager,
University College of Swansea

Adam Hilger Ltd, Bristol

British Library Cataloguing in Publication Data

Wood, Leslie B
The restoration of the tidal Thames.
1. Water—Thames River—Pollution
I. Title
628.1′688′422 TD425

ISBN 0-85274-447-1

Published by Adam Hilger Ltd, Techno House, Redcliffe Way, Bristol BS1 6NX.
The Adam Hilger book-publishing imprint is owned by The Institute of Physics.

Printed in Great Britain by Page Brothers (Norwich) Ltd.

To Kathleen

Preface

The work of the restoration of the Thames Tideway can be attributed to several organisations and to many people. The initiation of the work was due mainly to the Greater London Council and the Port of London Authority, with advice from central government, particularly the Water Pollution Research Laboratory. In the later stages the Thames Water Authority took over the responsibility from the GLC.

One Authority member stands out in his contribution to the work: Mr Peter Black was Chairman of the GLC and TWA during the decisive years. The London public are as much in his debt as the many officers who will remember his kind leadership and deem it a privilege to have served under him.

It should be remembered that while most of those involved in the later stages of the work have had the great satisfaction of seeing the results, there were many others who laid the foundations in the 1950s to whom this was denied.

In respect of this publication I would like to acknowledge the very great help of my ex-colleagues at the GLC and TWA, and to thank the latter body for permission to reproduce many of the diagrams. Especially, I wish to thank Michael Andrews whose devotion to the study of the flora and fauna of the estuary provided most of the biological information.

And last, but by no means least, I cannot let the very great help and support of my wife go unrecorded.

Leslie B Wood

Acknowledgments

The author wishes to express his deep gratitude to all those people and authorities who have provided information for this work, in particular to Thames Water Authority, without whose help the book could not have been produced.

The views expressed in this book are those of the author, and do not necessarily coincide with those of the bodies to which reference is made in the text.

Permission to reproduce the following illustrations is gratefully acknowledged.

Greater London Council:
  Curator of Maps and Prints: 31
  Director of Public Health Engineering: 94, 95, 96
HMSO (from *Greater London Drainage*): 47
Illustrated London News: 27, 28
Punch: 17, 18
Thames Board Limited: 77
Thames Water Authority: 1, 6, 10, 13, 14, 19, 22, 24, 25, 29, 30, 53, 57, 66, 67, 68, 69, 70, 71, 72, 73, 74, 82, 84, 106

Contents

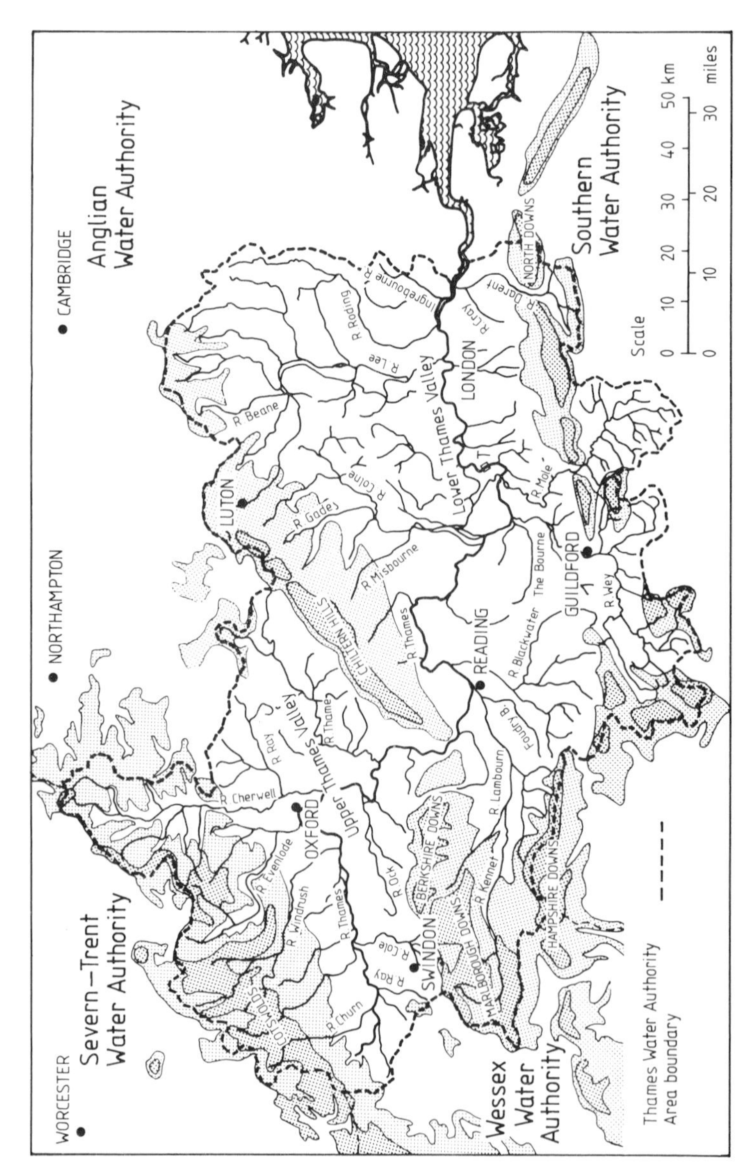

Figure 1. The River Thames (T = Teddington Weir). Courtesy of the Thames Water Authority.

Introduction

The Thames rises near Cirencester in Gloucestershire and flows as a freshwater river for 245 km to Teddington, which is 31.5 km above London Bridge (figure 1). At Teddington, a weir terminates the tidal effects of seaward penetration into the estuary. Below Teddington the tidal river flows through London—London Bridge is the point above or below which distances along the river are measured—and joins the sea. The exact line which separates the estuary from the sea has not been defined, but the seaward limit of the Port of London Authority (120 km below London Bridge) is probably as good as any.

The non-tidal (freshwater) river is entirely under the management of Thames Water Authority, the body which is also responsible for pollution control and water abstraction from the Thames and its tributaries as far downstream as to include the River Ingrebourne on the north bank, and the River Darent on the south. Thereafter, whilst Thames Water controls the river (Tideway) and tidal creeks down to a line about 83 km below London Bridge (Havengore Creek to Warden Point), the Anglian and Southern Water Authorities control the non-tidal tributaries to the north and south respectively.

The Tidal River Before 1800

Before the Ice Age it is doubtful whether the Thames existed in its present form. In the pre-glacial period the headwaters probably flowed from the Cotswolds, continuing past Oxford, along the present course of the River Ray (figure 2) to join what is now the River Great Ouse, and discharged into the Wash. The River Kennet flowed from the Marlborough Downs to Reading, and then along the present course of the Thames to its estuary in the North Sea. During the Ice Age, glacial pressure caused the ice to break through at the Goring Gap, to set the course of the Thames as we now know it. After the Ice Age the Thames would have been a freshwater river throughout its length (Kooijmans 1972), discharging together with other rivers of eastern England into a freshwater lake in what is now the North Sea. At this time sea level was so much lower than at present, that much of the bed of the southern and central North Sea would have been exposed as dry land. Eventually the fresh water cut through the chalk to form the Straits of Dover, whereupon the sea penetrated and rose in level to cut off the Thames from the other rivers. Since then the fauna of the Thames has been less diverse than that of the Ouse and Trent (Wheeler 1979).

The Development of London

London undoubtedly owes its existence to the River Thames. It is doubful whether there was a habitation there before Roman times. Possibly Julius Caesar noticed the terrain of the site of the present city and selected it as a base for a fortified settlement. He would have seen two gravel hills divided by a small stream (the Walbrook), with protection afforded on the west by the deep valley of the River Fleet, on the north by the fens of Finsbury, and on the east by swamps and the River Lee. The Thames itself provided protection to the south (Clayton 1964, Barton 1962). The Britons themselves may have developed settlements on each bank of the Thames between the times of Julius Caesar (55 BC) and Claudius Caesar (AD 43), but there seems to be little doubt that the Romans fortified the town during their occupation from AD 43–409 (figure 3). They built a wall around the town, dug a ditch for added protection, and may also have flooded Moorfields. The fortifications followed the line of gates which have left their names in Aldgate, Bishopsgate, Moorgate, Cripplegate, Aldersgate, Newgate and Ludgate (figure 4).

In a study of the risk of flooding of London, it is pointed out (Horner 1978) that south-eastern England is falling below mean sea level at the rate of 0.36 m per century. Based upon this, it can be seen from figure 5 that in Roman times the River Thames in London would hardly have been affected by the tides, and this is confirmed by the presence of the remains of yew trees in the peat of marshes near London. These trees are intolerant of water and will not live in saline conditions (Fitter 1945).

The river was probably wide and shallow, the areas of Westminster, Pimlico, Chelsea and Kensington being marshy. The Romans embanked the river and built a bridge where the first (stone) London Bridge would later be sited. The course of the river changed frequently; at one time it flowed directly from the present positions of Lambeth to Limehouse, and maps of

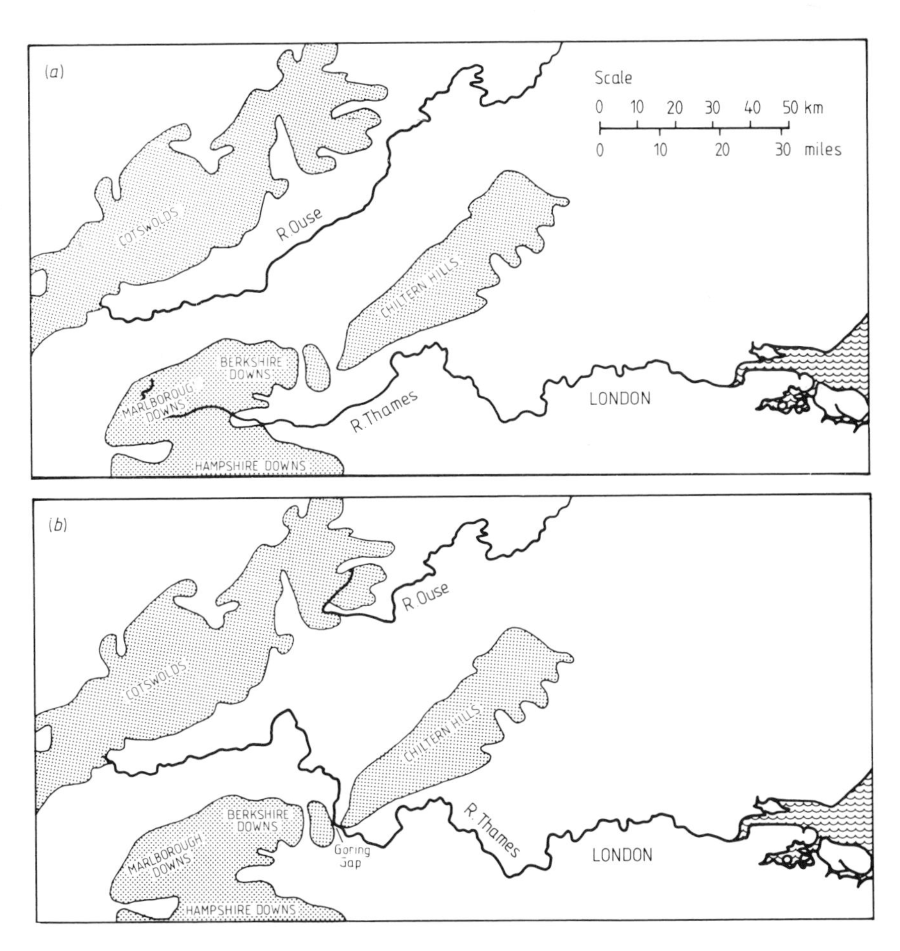

Figure 2. The courses of the Thames and Ouse (*a*) before the Ice Age, and (*b*) at present.

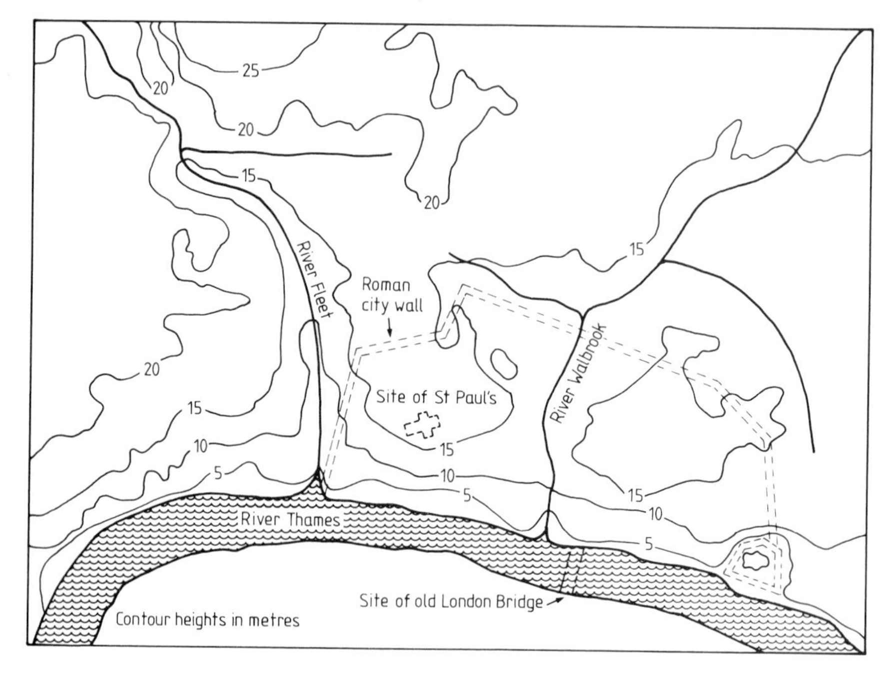

Figure 3. Pre-Roman London.

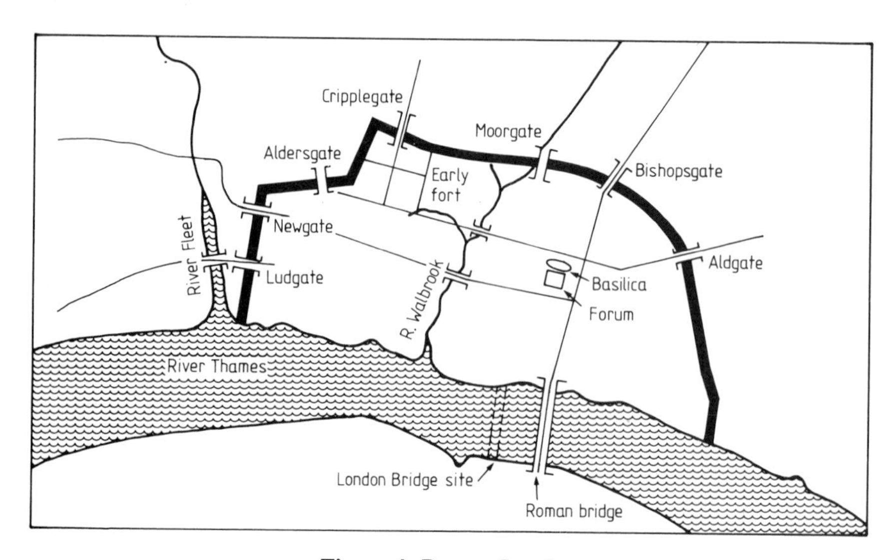

Figure 4. Roman London.

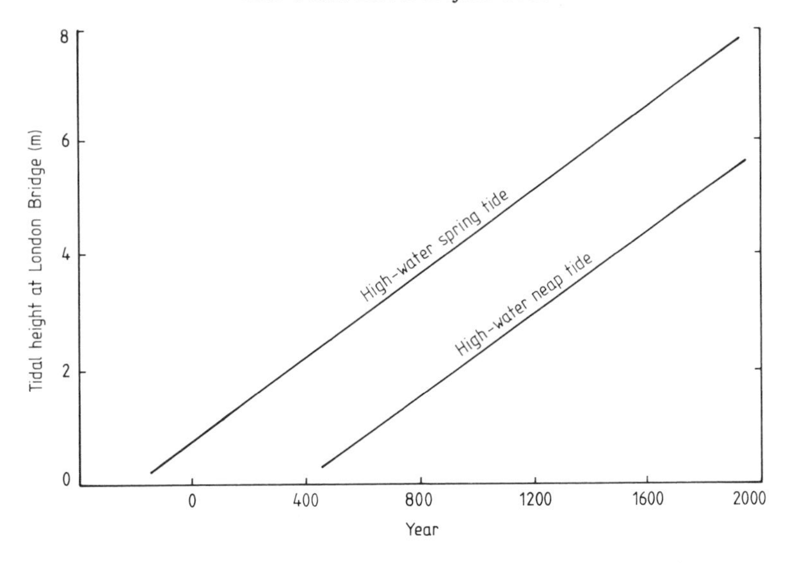

Figure 5. Variation in tidal height at London Bridge, 0–2000 AD.

the river, even up to the end of the eighteenth century (after which there was extensive embankment), show continual changes of course.

Many tributaries once flowed into the Thames in the present metropolitan area (figure 6). The Fleet was the largest, rising in two separate streams at Hampstead Heath and Highgate, and flowing to a confluence at Camden Town, which may have been the limit of navigation. The river continued to King's Cross, where its course can still be traced by the valley running to Holborn Viaduct (the site of Holborn Bridge), and thence to the Thames near Blackfriars Bridge. In Roman times it was thought to have been 60 feet wide at the Thames confluence, and upstream of Holborn Bridge it became the Hole Bourne, Oldbourne or Hollowburn, the names being derived from the steepness of its valley. It was also called the Turnmill Brook or River of Wells (Stow 1598)

The Walbrook rose in a swamp in Moorfields and flowed in a south-westerly direction to Bucklersbury. Its course lay along what is now the street called Walbrook, and beside the original Church of St Stephen's Walbrook, which was near the recently excavated site of the Temple of Mithras. Compared with the Fleet, the Walbrook was a small stream, only 12–14 feet wide and comparatively shallow, and joined the Thames near the site of the present Southwark Bridge.

The Tyburn was derived from two streams rising in Hampstead, and joining near Primrose Hill to flow through Regent's Park, dividing again into two streams before flowing into the Thames on either side of where Westminster Bridge now stands.

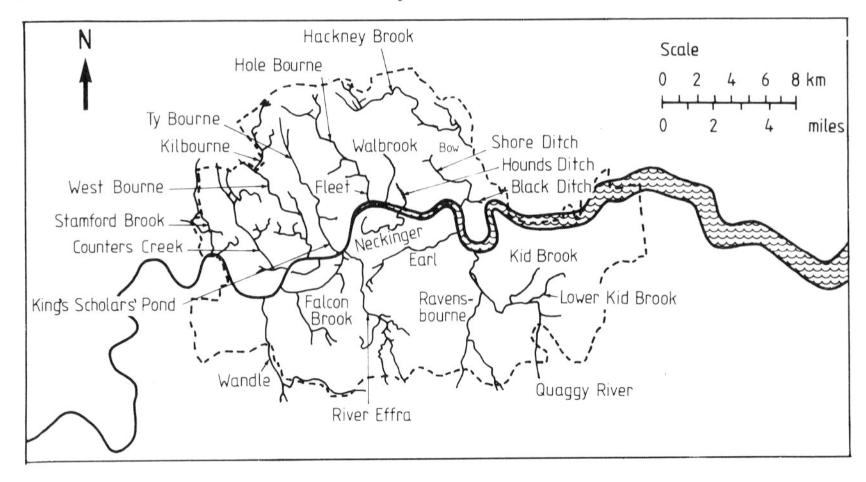

Figure 6. The 'lost' rivers of London.

On the south side of the Thames the main tributary was the Neckinger which took a semicircular course from Southwark to Bermondsey, joining the main river at St Saviour's Dock. In the course of time, all of these rivers and several others have been covered over and become part of the sewerage system, but throughout the city's history they have played a vital part in the livelihood of its citizens.

After the fall of the Roman Empire in the fifth century AD, London, like Rome itself, declined in population and for some time afterwards was probably

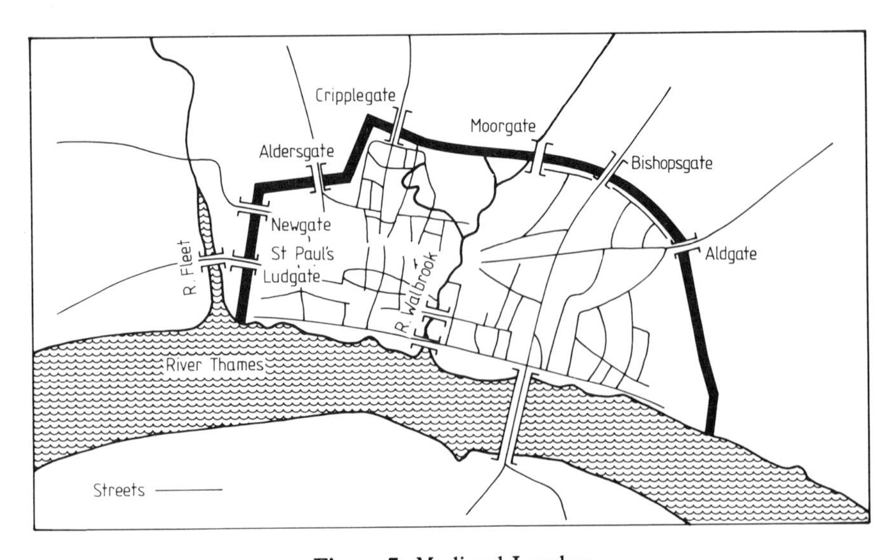

Figure 7. Medieval London.

little more than a Saxon settlement (figure 7). London was sacked by the Danes in 851 and 872, and when King Alfred rebuilt the town, the brick, stone and tiles of the Romans were replaced by wood and thatch, with the result that there were regular devastations by fire (in 982, 1077, 1087 and 1136).

Even up to the time of Henry II in the twelfth century, half of the city was still wasteland, and not until the reign of Henry III did it expand outside the Roman walls, when a charter in 1268 extended the jurisdiction of the city out to the suburbs. It is likely that it was not until the reign of Henry VII in the fifteenth century that the population again reached the level of that of Roman times, and thereafter it grew rapidly (figures 8 and 9).

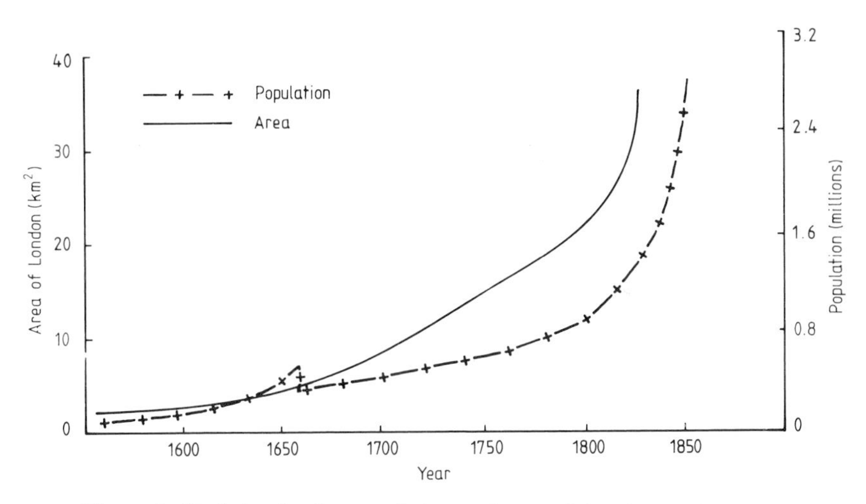

Figure 8. Variation in the population and area of London, 1550–1850.

Waste Disposal

In the earliest stages of urban development every house in London had its own plot of land where its domestic wastes could be deposited or buried. As the population increased, however, there was gradually less and less land available to each family, and a system of streets was laid out in the city. Thereafter, domestic refuse, cinders from foundries, offal from shambles, manure from stables, and like wastes, all found their way into kennels (gutters) which ran down the middle of the streets or along each side. Certainly until the early part of the fourteenth century the streets were neither paved nor cleansed, and where water ran through the foul streets to the rivers it became as polluting as sewage. Apart from street rubbish, each household developed its own heap composed of kitchen waste and manure from the many horses, pigs and other animals which were kept in the town. These middens were to some extent scavenged by the half-wild pigs which roamed freely about the streets until in 1292 they had become such a nuisance that men were appointed to kill all stray swine (Fitter 1945). The scavenging continued, however, by ravens and kites, and it became a capital offence to kill them. Refuse could not safely be burned because of the fire risk in the highly combustible medieval city.

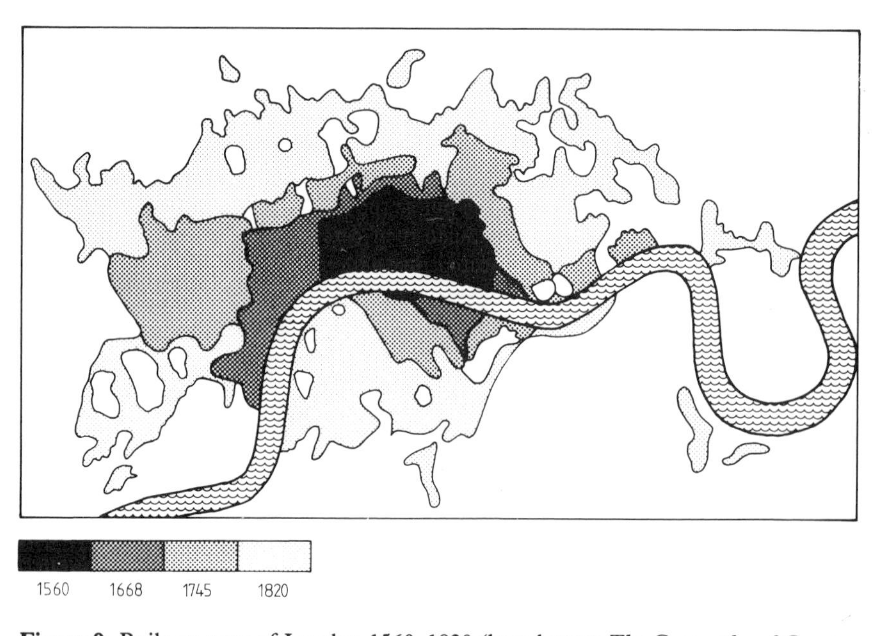

Figure 9. Built-up area of London 1560–1820 (based upon *The Geography of Greater London* ed. R Clayton, University of London Institute of Education, 1964).

In 1297 it became mandatory for every London householder to keep the frontiers of his tenement free from rubbish. In 1354 an order was made requiring the removal of rubbish every week, householders being responsible for putting out their accumulations on the appointed day. It was then collected by 'rakers' and put into tumbrils for carters to remove to be stored in laystalls. From these it was sold to farmers for agricultural use, or, in the case of riverside laystalls, to be shipped for dumping on the Essex marshes. In 1407 regulations required householders to store their refuse on their properties until the day of collection. The rakers were assistants to the Beadle of the Ward, whose responsibility it was to see that the regulations were carried out for cleansing kennels and laystalls; he had the authority to levy fines on householders who did not comply with the regulations.

Unfortunately the authorities found it very difficult to enforce the regulations. No lesser persons than John Shakespeare (father of the poet) and his friend and neighbour Adrian Quiney, both councillors of Stratford-upon-Avon, were fined for this offence in 1551. Edward III, when making a journey down the River Thames in 1357, found the stench so bad that he besought the Mayor and Sheriffs of London to forbid the throwing of rubbish in the Thames, Fleet, or into the fosses around the walls of the city. Householders living adjacent to the Walbrook had caused the flow to be stopped by throwing in rubbish, and in 1383 an ordinance was directed against them, requiring the piling or walling of the banks to ensure a free flow. An Act of 1388 prohibited the polluting of rivers, ditches and watercourses by deposits of filth, and ordained that all such refuse should be removed to appointed places before it became a nuisance.

The work of the rakers was later extended to clear large spaces such as the squares, but in 1628 Charles I complained of the fouling of the ways into the city. The Great Fire of 1666 appeared to have cleansed the streets, as complaints stopped for a time, but Lord Tyrconnel in 1741 considered that 'even a savage would look with amazement at the heaps of filth in London'.

Human wastes were collected in cesspools and removed by 'nightmen' for disposal to farmland to the north of the city, but no doubt cesspools sometimes overflowed and, just as today, doubtless there were some deliberately so constructed that they allowed some of the contents to leak into the ground to save the cost of frequent emptying. Neither is there any reason to assume that they were always emptied regularly.

Deterioration of Watercourses

Wherever there were stretches of flowing water, as in the 'lost' rivers, latrines were built over them and the contents discharged directly into them. This was permitted, provided that the watercourses were not used for the disposal of rubbish. Latrines caused problems at all times with their offensive odours, and during periods of high rainfall when they overflowed into cellars and houses. The supports of riverside latrines were used by fishermen to moor their boats, and the ropes not only tripped up passers-by but also weakened the latrine structures—with predictable results!

Eventually it became impossible to maintain the watercourses as free-flowing and unpolluted streams because of their abuse by the dumping of rubbish and discharges from the latrines. Also, siltation was a serious problem which could not be remedied, even by the high rates of flow associated with storms severe enough to cause loss of life and the destruction of buildings. In consequence, the use of these streams for water supply, navigation and recreation was lost. Ben Jonson (1616) referred to the River Fleet:

> In the first iawes appeared that ugly monster
> Ycleped Mud, which, when their oares did once stirre,
> Belch'd forth an ayre as hot as the muster
> Of all your night tubs, when the carts doe cluster,
> Who shall discharge first his merd-urinous load.

After the Great Fire Christopher Wren and Robert Hooke attempted to reconstruct the Fleet as an attractive watercourse, and devised and executed a plan to widen and deepen it. Unfortunately, however, the river continued to be used as a receptacle for rubbish, and silted so frequently that in 1733 it was covered over down to Fleet Bridge. In 1766 the remaining open stretches of the Fleet were covered, and in 1829 it disappeared completely under the Farringdon Road. For similar reasons the Walbrook was partly covered over in 1440; the work was extended in 1462, and by the end of the eighteenth century the river ran entirely underground. Many of the other London watercourses suffered a similar fate.

Drainage

As building developed in the city it provided an impermeable surface which prevented natural drainage by way of water seepage into the ground, and thence by the tributaries, to the Thames. The provision of kennels in the streets alleviated the situation to some extent, but severe flooding still occurred at regular intervals. The Fleet was particularly susceptible; in 1317 when in spate it damaged bridges as well as houses and mills, and several people were drowned.

An Act passed under Henry VI appointed a Commission of Sewers, and the Commissioners had the responsibility of enquiring into the risk of flooding. The Bill of Sewers 1531 in the reign of Henry VIII consolidated previous legislation and established a system which was to operate for the next 300 years. The Commissioners appointed juries to investigate specific areas and to report on what improvements were necessary, whereupon the Commission set out its recommendations. The Commissioners had great authority, even to the extent that in 1663 they directed Charles II to bear half of the cost of clearing the rubbish from a sluice by his bowling green. Also, in 1768 they permitted George III, at his own cost, to alter the course of the King's Scholars' Pond sewer from under the Queen's Palace at St James.

The sewerage system was designed to carry away only surface runoff and until 1815 it was a penal offence to discharge sewage or other offensive matter into it. Nevertheless, bearing in mind the difficulties of controlling the dumping of refuse in the streets, it is probable that at times these sewers were highly polluted.

Water Supplies

In early times the citizens of London obtained their water directly from the Thames and its tributaries, or from the springs (also called fountains or wells) in the higher ground to the north and west of the city. In 1180, during the reign of Henry II, Fitzstephen (originally personal secretary to Thomas Becket) wrote: 'towards the north [of the City] arise excellent springs at a small distance whose waters are sweet, salubrious, clear . . .' Holywell, Clerkenwell (Clerke's Well), and St Clement's Well were those most frequently used, but in 1603 Stow referred to other springs near Clerkenwell, such as Skinner's, Fag's, Tode, Loder's Wells and Radwell. Pools were often used, such as Horsepool near West Smithfield, and another near to the Church of St Giles Cripplegate. The former must have been of some size, since in 1244 Anne of Lodburie was drowned therein.

Stow considered that these sources of potable water must have been satisfactory until the beginning of the fourteenth century, after which time, because of problems of the city's growth, the resources had deteriorated both in quality as well as quantity, and it was necessary to seek supplies from farther afield. Problems had also arisen in obtaining supplies from the Thames. In 1342, the Mayor, aldermen and commonalty of the city had received numerous complaints that the lanes to the River Thames had been closed, and that no one was allowed to pass without paying a duty. An inquisition was made and 'divers persons of several wards' were sworn to make diligent enquiry into these grievances.

In 1439 Henry VI confirmed the granting to the citizens of London by the Abbot of Westminster, a head of water 143 yards long and 5 yards wide, and all the springs in the Manor of Paddington, for a yearly rent of two peppercorns in perpetuity.

By the time of Stow, at the end of the sixteenth century, a system of conduits had been developed to replace the local supplies. Water was conveyed by leaden pipes to outlets called conduits, from which water supplies could be obtained either directly, or from people engaged in selling it from 'tankards' which contained about three gallons, and were carried on the head or shoulders by men and women†. The first conduit was laid in Westcheap, started in 1235 and completed about 50 years later. In 1236 two merchants of Amiens, Nele and Corby, were granted by the Mayor the privilege of landing and storing woad, for which they paid a yearly sum of 50 marks, and a donation of 100 pounds towards the expense of conveying water from 'Tyborne', then a village, to the city. In the fifteenth century pipes were laid to Cheapside and later to Fleet Street, Aldermanbury and Cripplegate. By the middle of the sixteenth century there were conduits in many parts of London (figure 10), the expense of their construction and maintenance being met by various prominent citizens and civic dignitaries. The conduits were often used for displaying 'moral' sentences such as the one at Cornhill.

> Bread earned with honest labouring hands,
> Tastes better than the fruite of ill-got lands.

After 1568 new conduits were built to supply water from the Thames. Mechanical assistance was being developed for obtaining water at this time, the conduit at Dowgate being supplied from a horse-driven machine in Cosen Lane.

† They were called 'cobs' after Cob Court which led down to the river where most of them lived.

Figure 10. A conduit in London.

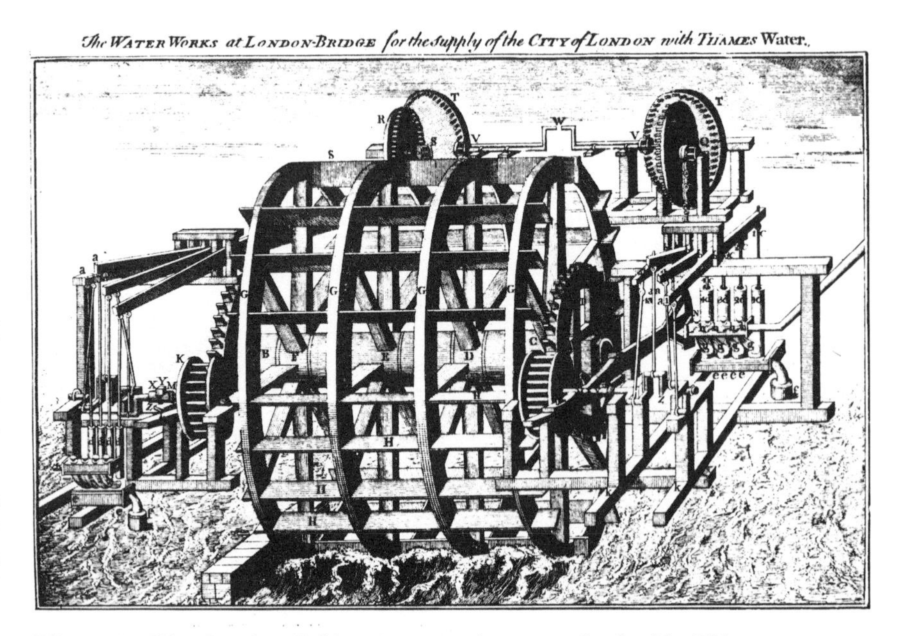

Figure 11. The London Bridge water works waterwheel with lifting apparatus, designed by Peter Morris (from the *Universal Magazine*, 1749).

In 1581 a Dutchman, Peter Morris, was granted by the city a 500-year lease of the first arch of London Bridge for the installation of his famous waterwheel (figure 11), for which he paid a yearly sum of ten shillings. His device confined the flow of the river to such an extent that a great velocity was produced by the ebb and flow of the tide, and when the installation was tested, it threw a jet of water over the steeple of the Church of St Magnus the Martyr. Morris supplied conduits at the Churches of St Mary Magdalene and St Nicholas Cole Abbey near Old Fish Street. The granting of a lease to Morris created a precedent, in that it empowered a private individual to build a waterworks of his own, with the object of supplying private houses with piped water from street mains (Berry 1957). Bevis Bulmer in 1594 erected machinery at Broken Wharf to supply the western part of the City.

In 1570, in the thirteenth year of the reign of Elizabeth I an Act of Parliament was passed for bringing a river from 'any part of Middlesex or Hertfordshire to the Metropolis'. Although this was never acted upon, in 1607 a similar Act was passed, which resulted in a plan by Colthurst in 1609 for bringing water from springs at Amwell in Hertfordshire to Islington. The work was eventually carried out by Sir Hugh Myddleton (figure 12), citizen and goldsmith of London and Member of Parliament for Denbigh.

Figure 12. Sir Hugh Myddleton (from the original engraving in Goldsmith's Hall, London).

In order for him to complete the work in the four years allotted to him by the agreement, Myddleton employed some 200 labourers in addition to many skilled men, although by the time that the 'New River' had reached Enfield, his financial resources were exhausted. He sought help from James I, who paid for half of the whole cost and provided wayleaves through Royal manors and lands. The work was eventually completed so that on Michaelmas Day 1613, water gushed into the Round Pond near Sadler's Wells from the Springs of Amwell, having been brought by a winding course of nearly 40 miles to supply the surrounding parts of the city (figure 13).

During the eighteenth and nineteenth centuries, following the development of pumping machinery by James Watt, several companies were established which drew their water supplies from the tidal Thames, as shown in figure 14 and listed in table 1.

Table 1. Companies which drew water from the tidal Thames early in the nineteenth century.

Company	Source	Year founded	No. of houses supplied	Gallons/day (average)
New River	Amwell Springs	1619	73 212	241
Chelsea	Thames N side	1723	13 891	168
West Middlesex	Thames N side	1806	16 000	185
Grand Junction	Thames N side	1811	11 141	350
East London	Thames N side	1807	46 421	120
South London	Thames S side	1805	12 046	100
Lambeth	Thames S side	1785	16 682	124
Southwark	Thames S side	1845†	7100	156

† An amalgam of two companies: Southwark founded in 1760, and Vauxhall in 1805.

Fish in London's Rivers

Little appears to have been recorded about fish in the rivers which have now become some of London's sewers. However, in the days of Fitzstephen in the twelfth century, the tidal River Thames abounded with fish and the citizens of London found freshwater fish (especially eels) to be a very important part of their diet. The city fathers had jurisdiction over the river from Staines to Yantlet (near Canvey Island), and records show ordinances of the reign of Richard I for the removal of weirs which hindered navigation and destroyed fish fry. In 1285 Edward I, by the Statute of Westminster,

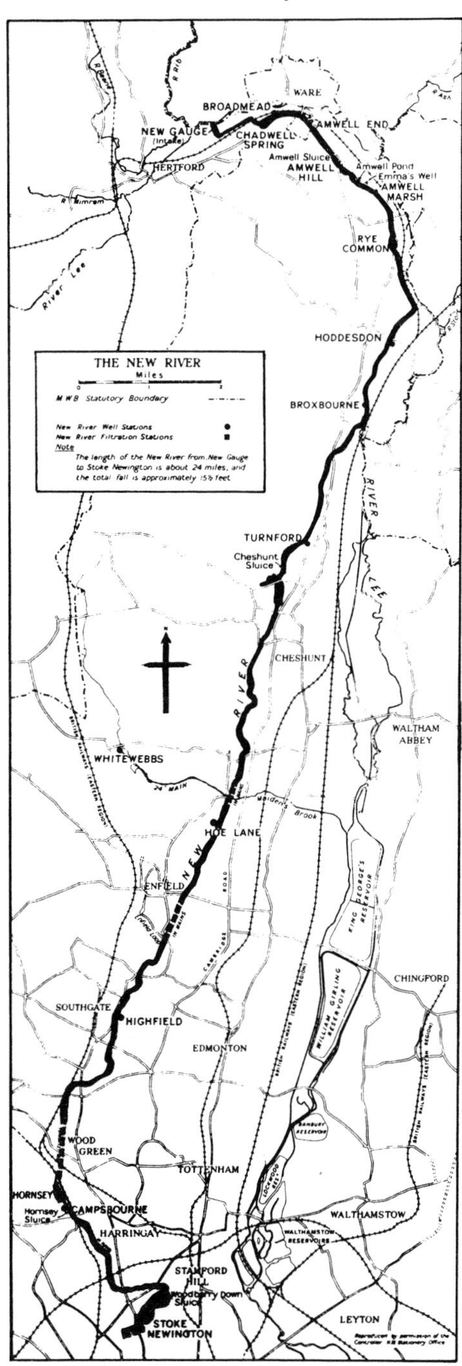

Figure 13. The 'New' River.

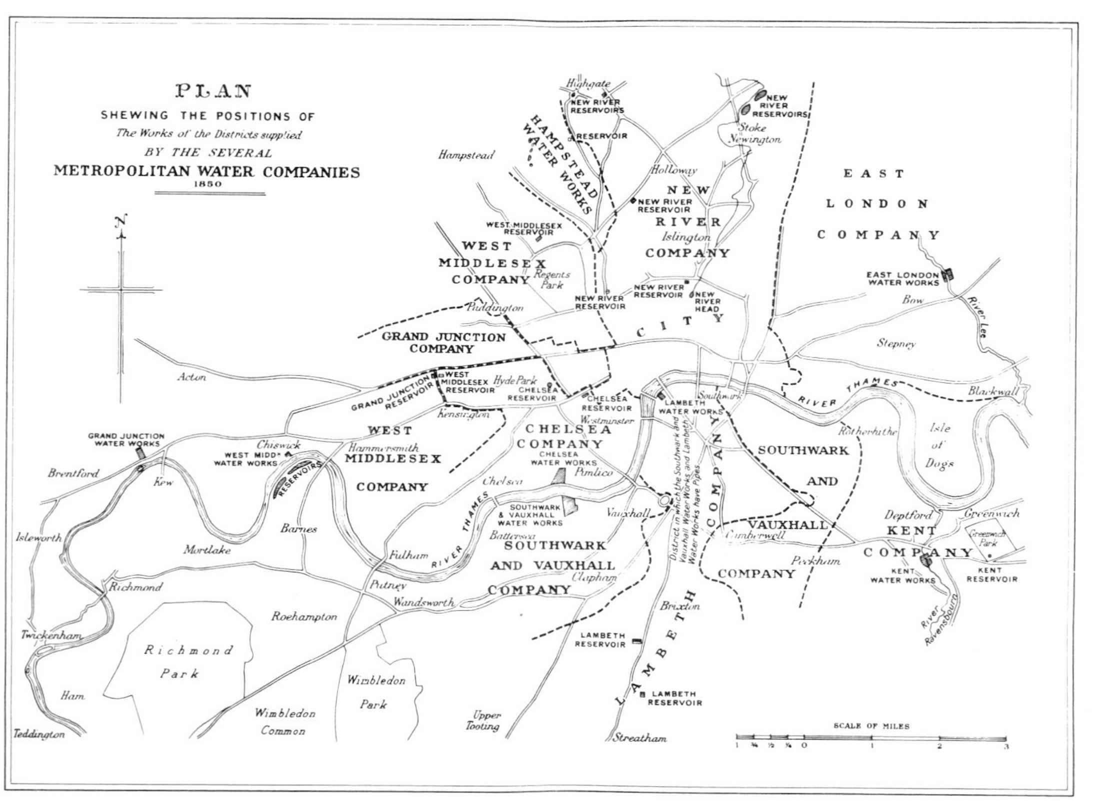

Figure 14. Map showing areas of the metropolitan water companies, 1850.

introduced a close season for salmon fishing, and Richard II, in a further Act, protected salmon and lampreys in the river. There are several historical instances relating to the fish of the Thames; Henry III was presented with a white bear which was kept in the Tower of London and taken out by a keeper to fish for salmon. The pike, eels and pickled sturgeon presented to Edward III on his marriage to Philippa in 1328 probably came from the Thames. In 1457 (Cornish 1902) it was recorded that four 'great fysshes' were caught between London and Erith. They were not in fact fish, being two whales, a walrus and an alleged 'swordfish' which was probably a narwhal, but no doubt they were consumed with readiness by the citizens.

There is little doubt that in Tudor and Stuart times fish were taken in abundance from the river and there were elaborate regulations to prevent overfishing. An Act of 1630 established a specific close season for salmon fishing in the Thames, which leads us to suppose that the stock was of sufficient size to warrant protection, although it is doubtful whether the numbers approached the magnitude of those of the rivers of the north and west of England. On 7 June 1749 (Williams 1946) 47 salmon were caught below Richmond in two draughts of the net, which lowered the price of these fish at Billingsgate from 1s 0d to 6d per pound. Again, in July 1766, 130 salmon were taken from the Thames in one day, and were sent to market. In 1810, a Mr Goldham of Billingsgate reported that ten salmon and 3000 smelt were taken in one haul from the river near Wandsworth, and 3000 Thames salmon were taken to market in that year (Welch 1894).

Robert Binell, a water bailiff of the city, said in 1757: 'There is no river in all Europe that is a better nourisher of its fish and a more speedy breeder, particularly of flounder, than is the Thames.' In the year 1819, it was recorded that the following freshwater fish species could be found in the Thames in Middlesex: salmon, trout, grayling, perch, carp, tench, roach, dace, gudgeon, pike, eels and lampreys. Salt-water fish also found included sole, plaice, skate, halibut, haddock, oysters, mussels and prawns (Fitter 1945). Hofland in 1848 referred also barbel, chub and flounders as being plentiful, and stated that smelt could be caught at London Bridge. Table 2 shows catches of salmon at Boulter's Lock from 1794–1821. As late as 1828 there were 400 fishermen earning their livelihoods from the river between Deptford and London.

Condition of Rivers in London at the End of the Eighteenth Century

By the end of the eighteenth century most of the tributaries of the tidal Thames in London had become so polluted that they had been covered and so formed part of the sewerage system. Despite the loads discharged into the Tideway from these sewers, however, the condition of the river must still have been satisfactory in London. Salmon and other fish were still being taken in plentiful numbers, so that the dissolved oxygen concentration must have been above 35% saturation, even under the worst conditions. There would have been more diluting flow from the Upper Thames than there is at present, since the abstraction from that part of the river would have been much lower because of the smaller population drawing potable water supplies from it. Seven water companies drew their supplies from the river below Teddington in the early 1800s, without undue complaint from their customers.

Table 2. Catches of salmon at Boulter's Lock, Taplow, 1794–1821 (from Report of the Thames Migratory Fish Committee 1978, Thames Water Authority).

Year	Number	Total weight (lb)	Five-year running average number
1794	15	248	—
1795	19	168	—
1796	18	328	—
1797	37	670	—
1798	16	317	21
1799	36	507	25
1800	29	388	27
1801	66	1124	37
1802	18	297	33
1803	20	374	34
1804	62	943	39
1805	7	116	35
1806	12	245	24
1807	16	253	23
1808	5	88	20
1809	8	116	10
1810	4	70	9
1811	16	182	10
1812	18	224	10
1813	14	220	12
1814	13	98	13
1815	4	52	13
1816	14	179	13
1817	5	76	10
1818	4	49	16
1819	5	84	6
1820	0	0	5
1821	2	31	3

It is not possible, of course, to state precisely the condition of the tidal river, since systematic analyses were not made until 1882, but bearing in mind such records that do exist, however, it is probable that its quality was not greatly different from that of today.

The Deterioration of the River, 1800–50

During the first half of the nineteenth century London was undergoing considerable change. The British Empire was expanding rapidly, with London as the principal city dealing with a very substantial proportion of the commerce of an Empire which, at its peak, handled one-third of world trade.

The Industrial Revolution

The Industrial Revolution changed the craftsman's life in that instead of working in his own home, he now had to go to work in a factory. New industries were being established in the environs of the metropolis, producing effluents containing artefacts not always amenable to biodegradation by the natural processes in rivers.

Of all the wastes from the numerous metropolitan industries, that from chemical and gas manufacture was probably the most polluting. The Gas Light and Coke Company was set up by Act of Parliament in 1810, and soon thereafter other companies were established in London for gas manufacture, and it became necessary to provide separate districts for each by the Metropolitan Gas Act of 1860. The by-products which later provided a source of valuable chemicals were at the time the cause of considerable pollution in the tidal river. The worst pollutants were phenols and ammonia, both of which are very toxic to aquatic life and have high oxygen demands.

The population of Britain increased rapidly in the nineteenth century. The death rate had fallen, largely because of the improved living conditions and medical treatment available to the poorer classes, but in no small measure it was also due to the change from the excessive consumption of spirits (gin) to the drinking of tea (Trevelyan 1944). Figure 15 shows how the population of London grew from just over one million in 1801 to about 2.75 millions in 1851, an increase which inevitably led to the generation of more wastes, and to the discharge of a greater pollution load to the Tideway.

Moreover, railways had been constructed; the rigorous twelve-hour stage-coach journey from Birmingham to London had been reduced to a comparatively easy train journey of little more than two hours. Thus not

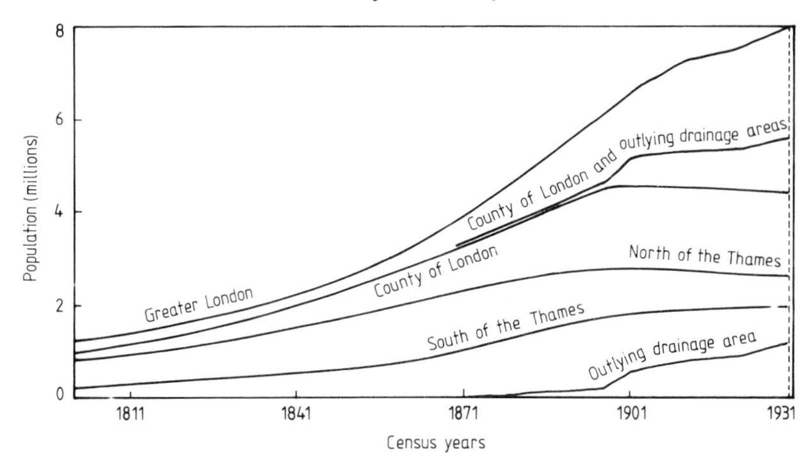

Figure 15. The population of London, 1801–1930.

only did the commuting public swell the resident population, but there was also considerable movement into the capital by visitors from other cities.

The Disposal of Human Wastes

The development which was to have the greatest effect on the deterioration of the Tideway, however, was the introduction on a large scale of water-carriage of human wastes by the use of the water closet. The invention can probably be attributed to Sir John Harington (Harington 1596), although his rather crude apparatus (figure 16) did not come immediately into general use; it was not until Joseph Bramah produced a much improved version 200 years later that it became more widespread. Between 1778 and 1797 Bramah supplied more than 6000 water closets and they became widely used in London after 1830.

Human wastes had previously been disposed of in outdoor privies discharging to cesspools, the contents of which were removed by 'nightmen' or scavengers since the connection of these systems to the street sewers was forbidden. The indoor water closet must have made a considerable contribution to creature comforts: Lord Byron was once ejected from Long's Hotel in Bond Street when on a cold, wet night he preferred to use the hall rather than the outside privy!

In 1800 there were about 200 000 houses in London, most of which had privies with cesspools. By 1850, when the number of houses had risen to 300 000, legislation required that any new properties were to have water closets and although before 1815 it had been an offence to do so, in that year it became mandatory to connect cesspools to the sewers. This legislation, introduced by the First Metropolitan Commissioners of Sewers (p 28), had the disastrous effect of causing the entire contents of water closets to be discharged amost directly into the Thames in the heart of the metropolis, by way of what were designed to be sewers for surface water only. As a result the river became putrid and the mud-banks covered with sewage solids.

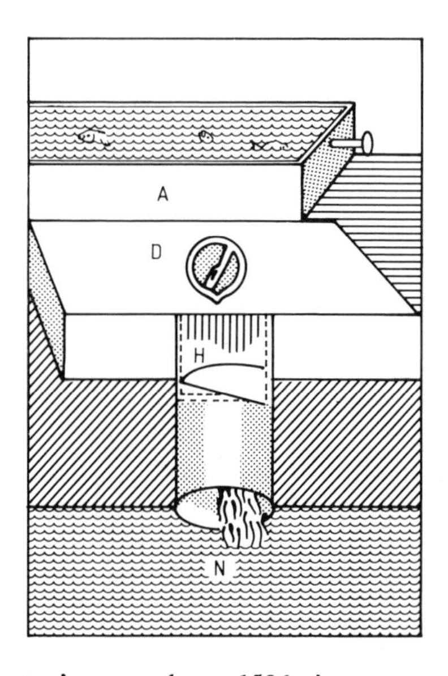

Figure 16. Harington's water closet, 1596. A, water reservoir; D, seat; H, receptacle for excreta; N, vault for receiving waste. It was recommended that six inches of water should be left in chamber H, and that it be flushed at noon and night.

Condition of the River in the Mid-nineteenth Century

The state of the river was eloquently described in *Punch* in 1842, as follows:

> Lambeth pours forth a rich amalgam from the yards of knackers and bone grinders. Horseferry liberally gives up all its dead dogs. Westminster empties its treasures into the mighty stream by means of a common sewer of uncommon dimensions, the Fleet ditch bears its inky currents in concentrated essences of Clerkenwell, Fieldlane, Smithfield, Cowcross, St Giles and the surrounding districts. The fluids of Whitechapel slaughterhouses call in transit for the unlovely contribution of Houndsditch, Ratcliff Highway, Bevis Marks and Goodman's Fields and thus, richly laden, they pour their delicious slime into the Thames by means of Tower ditch.

By the 1850s the smell of the river was notorious. It was said that when people on the river banks saw a paddle steamer approaching, they would flee in anticipation of the smell it would produce by stirring up the water and releasing the dissolved hydrogen sulphide. At first the smell was worst in the region of Bermondsey and Southwark, where the poor and uninfluential working population had to endure it in silence. Michael Faraday commented on the condition of the Thames in *The Times* of 9 July 1855, as shown below.

Extract from the TIMES, *July 9th,* 1855.

"THE STATE OF THE THAMES.

"To THE EDITOR OF THE TIMES.

"SIR,—I traversed this day, by steamboat, the space between London and Hungerford Bridges, between half-past 1 and 2 o'clock; it was low water, and I think the tide must have been near the turn. The appearance and the smell of the water forced themselves at once on my attention. The whole of the river was an opaque, pale brown fluid. In order to test the degree of opacity, I tore up some white cards into pieces, moistened them so as to make them sink easily below the surface, and then dropped some of these pieces into the water at every pier the boat came to; before they had sunk an inch below the surface they were indistinguishable, though the sun shone brightly at the time, and when the pieces fell edgeways the lower part was hidden from sight before the upper part was under water. This happened at St. Paul's Wharf, Blackfriars Bridge, Temple Wharf, Southwark Bridge, and Hungerford Bridge; and I have no doubt would have occurred further up and down the river. Near the bridges the feculence rolled up in clouds so dense that they were visible at the surface, even in water of this kind.

"The smell was very bad, and common to the whole of the water; it was the same as that which now comes up from the gully-holes in the streets; the whole river was for the time a real sewer. Having just returned from out of the country air, I was, perhaps, more affected by it than others; but I do not think I could have gone on to Lambeth or Chelsea, and I was glad to enter the streets for an atmosphere which, except near the sink-holes, I found much sweeter than that on the river. I have thought it a duty to record these facts, that they may be brought to the attention of those who exercise power or have responsibility in relation to the condition of our river; there is nothing figurative in the words I have employed, or any approach to exaggeration; they are the simple truth. If there be sufficient authority to remove a putrescent pond from the neighbourhood of a few simple dwellings, surely the river, which flows so many miles through London, ought not to be allowed to become a fermenting sewer.

"The condition in which I saw the Thames may, perhaps, be considered as exceptional, but it ought to be an impossible state, instead of which, I fear, it is rapidly becoming the general condition. If we neglect this subject, we cannot expect to do so with impunity, nor ought we to be surprised if, ere many years are over, a hot season gives us sad proof of the folly of our carelessness.

"I am, Sir, your obedient Servant,

"M. FARADAY."

"Royal Institution,

"*July 7th,* 1855."

The 'Year of the Great Stink'

The year of 1858 began fairly dry, and by June the temperature had reached 95°F. Consequently, as Faraday had predicted, the river stank thoughout London from east to west, so that its polluted condition was brought physically to the attention of the influential, and 1858 became known as the 'Year of the Great Stink'.

In June of that year, Queen Victoria, the King of the Belgians, the Duke of Brabant, and several other British and foreign dignitaries were crossing the Thames from Deptford to inspect the new ship (later to be known as the

Great Eastern) when the Queen was compelled to hold her bouquet to her face to avoid the smell, and the whole party were glad to get ashore. In the same year, the smell in the neighbourhood of the Houses of Parliament caused so much unpleasantness to Members that the windows were hung with sheets soaked in disinfectant to counteract the smell and to allow the business of the House to proceed. Without doubt, this must have prompted the more expeditious transactions of the legislative procedures for abatement of the Thames pollution then in hand!

> Filthy river, filthy river,
> Foul from London to the Nore,
> What art thou but one vast gutter,
> One tremendous common shore!
> All beside thy sludgy waters,
> All beside thy reeking ooze,
> Christian folk inhale mephitis,
> Which thy bubbly bosom brews.
>
> *Punch* (1858) vol. xv, p161.

THE "SILENT HIGHWAY"-MAN.

"Your MONEY or your LIFE!"

Figure 17. A *Punch* cartoon of 1842.

The Cholera Epidemics

But far more serious than the visual and olfactory nuisance from the polluted river was the risk of the transmission of disease as a result of the contamination of water supplies drawn directly from the river. The citizens of London, particularly the poor, were at great risk from waterborne infection at this time. The poor obtained their water either from local pumps drawing from shallow wells (which in some cases drained cesspools or burial grounds), or from stand-pipes receiving water from the Thames, where the potable water intakes and sewerage outlets were often in close proximity. In most cases water was merely supplied and not purified, and as an inevitable result London experienced its first epidemic of Asiatic cholera in 1831–32, followed by similar serious outbreaks in 1848–49, 1853–54, 1865–66, and a minor

FARADAY GIVING HIS CARD TO FATHER THAMES;

And we hope the Dirty Fellow will consult the learned **Professor.**

Figure 18. 1858—the Year of the Great Stink.

recurrence in 1871. In 1849 there were 18 036 deaths in London, and in 1854 nearly 20 000. Of the 896 residents of a small area of Soho, 500 people died in the first ten days of September 1854 (Kekwick 1965). Dr John Snow carried out some research into the occurrence of cholera (see table 3), and recognised a clear connection between the disease and the polluted water sources, despite the physical clarity of the latter. Pressure of public opinion soon brought about an enquiry into public health administration. In particular, in 1834 Edwin Chadwick was appointed secretary to the Poor Law Commission which, through its classic report of 1842 (Chadwick 1842) greatly influenced the subsequent review of the laws of public health.

Table 3. Dr Snow's survey of deaths from cholera.

District	Density of cholera	Water supply company	Water source
Southwark	71 in 10 000	Southwark	Tidal Thames
Lambeth	5 in 10 000	Lambeth	Thames above Teddington

In 1850 the General Board of Health recommended the consolidation of the management of the various water supply works, as well as drainage and sewerage works. It also recommended that a constant water supply should be introduced, to replace the intermittent supply which had led consumers to store water in large, unhygienic receptacles for use in times when the supply was stopped. In 1852 the Metropolis Water Act was passed, which prohibited the abstraction of water from the tidal reaches of the Thames below Teddington Weir.

In such a situation as was presented in London at the time, inevitably some interests suffered in the remedying of others. Chadwick in this case firmly supported water carriage of wastes, and no doubt influenced the widespread use of street sewers for disposal, with the result that the self-purifying capacity of the Tideway was exceeded by the extra pollution load, and it became a putrid, noisome and dead river in London.

The First Restoration, 1850–1900

Coordination of Sewerage

By 1850 in the 150 km^2 of London there were about 150 km of closed, and 333 km of open drainage ditches, discharging to 70 outlets in the metropolitan area (Mayhew 1968), as shown in figure 19. Drainage was controlled by eight independent Commissions of Sewers, namely:

The City of London
Westminster and part of Middlesex
Holborn and Finsbury
The Tower Hamlets
St Katherine
Poplar and Stockwell
Greenwich
Surrey and Kent

Each commission had sole powers in respect of its own operations, including the selection of sizes of drains, their inclinations and methods of construction.

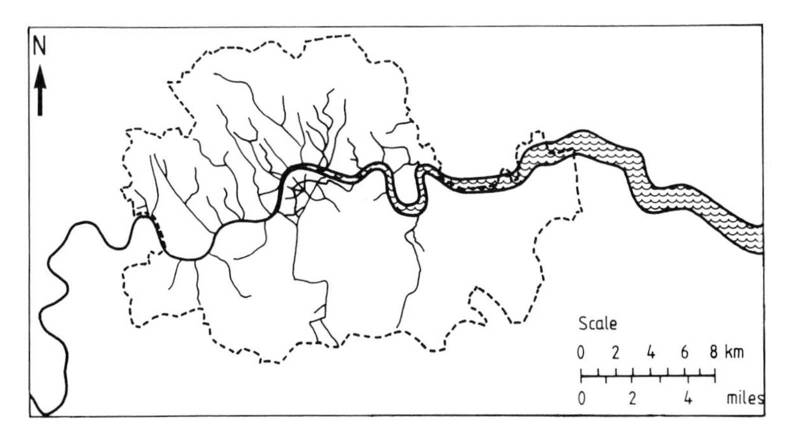

Figure 19. Sewerage of London, 1850.

Thus at district boundaries, sewers might be of different sizes, shapes and levels, and in some cases large sewers discharged in smaller ones. It soon became clear that coordination was necessary to remedy the defective London drainage system, and in 1848 an 'Act to Consolidate the Metropolitan Commissions of Sewers' was passed, which enabled Her Majesty to unite the previous eight commissions into one, called 'The Metropolitan Commission of Sewers', which had jurisdiction over

> . . . places or parts in the counties of Middlesex, Surrey, Essex and Kent, or any of them, not more than twelve miles distant in a straight line from St Paul's Cathedral but not being within the City of London or the liberties thereof.

The Commission was required to ensure that every house in London was provided with a water closet or privy, and to make a complete survey of the sewerage system of London.

The Metropolitan Commission of Sewers

The First Metropolitan Commission of Sewers was set up in 1847 and began an improvement of domestic drainage and the abolition of cesspools, the latter operation causing considerable pollution to the tidal Thames, to which reference has already been made.

The Second Commission superseded the First in 1849, and it considered that sewage should be kept out of the Thames completely. Two schemes were put forward: the engineer Austin suggested conveying sewage to large tanks or sumps to be taken to land for agricultural purposes. Phillips, the chief surveyor, suggested using a system for intercepting the sewers and conveying sewage to the Thames downstream of London. Such a plan had already been suggested in 1834 by John Martin, who was at that time an artist well known for his biblical paintings such as 'Belshazzar's Feast', 'The Great Day of Wrath' and 'The Deluge'. Martin's plan had been to embank both sides of the Thames in London from Westminster to Blackfriars, and to construct an intercepting sewer in each embankment to collect the sewage and to discharge it to the river downstream, where the population was sparser. A committee sat to examine the plan, under the chairmanship of the Earl of Euston, and included the scientists Wheatstone and Faraday. The committee gave its approval†, but the two government select committees to whom it was subsequently put failed to act upon it. However, the plan was taken up again in 1845 by Thomas Wicksteed (engineer to the East London Waterworks Company), who proposed that the intercepting sewers should be extended

† Report of Committee appointed to Take into Consideration Mr Martin's Plan for Rescuing the River Thames from Every Species of Pollution, London, 23rd April, 1836.

to outfalls at Barking Creek on the north bank, and at Greenwich Marshes on the south bank, and the sewage deodorised prior to discharge. The Second Commission was unable to decide whether to support the rival plans of Austin or Phillips, and so advertised for competitive plans for a complete system of drainage for London. Even though 116 designs were received, the Commission was unable to reach a decision, and resigned.

The Third Metropolitan Commission appointed later in 1849, which had amongst its members several eminent engineers, including the railway engineer Robert Stephenson, deliberated upon the plans. Much merit was found in the plan by McLean and Stileman (figure 20), which proposed carrying intercepting sewers to outfalls on the sea, whereby the sewage could be used for reclaiming waste land on the Essex coast. In 1850, however, the Commission concluded that none of the schemes could be recommended, and decided to appoint its own engineer to prepare a comprehensive plan for main drainage of the metropolis, taking into account all previous suggestions and information, and Mr Frank Forster was appointed. His plan for the south side of the river was to collect sewage at the River Ravensbourne, and to lift it 25 feet so that it would flow to Woolwich Marshes, where a second pumping station would raise it to a reservoir for discharge. For the north side, he proposed a high- and low-level intercepting sewer to meet at the River Lee, where the low-level sewage would be pumped to join that from the high level, to flow to a reservoir near Gallion's Reach at Woolwich. From both reservoirs the sewage would be discharged at the time during the tidal cycle which was such that the least flow would return upriver. The plan involved the construction of 63 km of sewers on the north and 22 km on the south, at an estimated cost of £1.5 million. The money was not forthcoming, and in 1852 the Third Commission was superseded by the Fourth, which considered Forster's plan. However, one of its members, Captain Vetch, put forward an alternative and there was a further division of opinion. Forster resigned and died shortly afterwards.

The Fourth Metropolitan Commission was succeeded by the Fifth in 1852, the members of which appointed Joseph Bazalgette (figure 21) as engineer and instructed him, in conjunction with Mr Haywood, to prepare a scheme for the drainage of London north of the Thames. Their plan was a modified version of Forster's. It was at this time that the General Board of Health proposed a plan for remodelling the London drainage system entirely, on the basis of a 'separate system', whereby excess rainwater (which, being relatively unpolluted, could be discharged directly to rivers), was separated from the 'foul' sewage by providing independent channels for the removal of each. The Secretary of State for the Home Department supported this plan, and communicated his opinion to the Fifth Commission, who did not agree with him and also resigned.

At the end of 1854 a Sixth Commission was appointed who invited new plans, considered several of them but produced no solution. Meanwhile, as

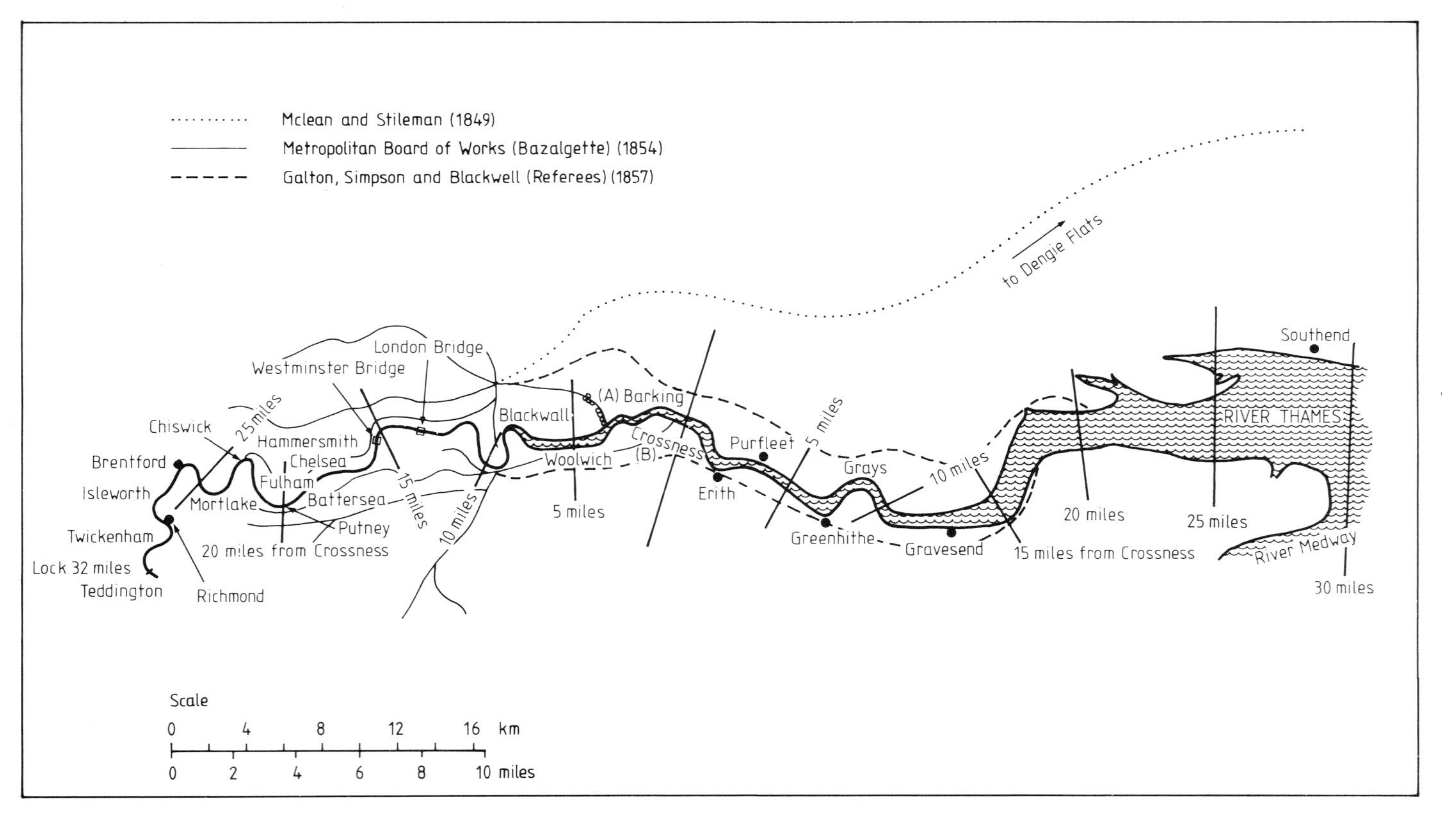

Figure 20. Alternative plan of Galton, Simpson and Blackwell.

a result of the abolition of 200 000 cesspools by the First Commission, the banks of the Thames had become covered by sewage solids, the smell was putrid, and there had already been three severe cholera epidemics.

Figure 21. Sir Joseph Bazalgette.

The Metropolitan Board of Works

It was obvious that there was a need to provide an effective body capable of planning and administering, *inter alia*, the work of London's drainage, and a Bill introduced by Sir Benjamin Hall (First Commissioner of Works) became 'An Act for the Better Local Management of the Metropolis 1855,' which received Royal assent in 1856. The Act provided for the creation of district boards to control local drainage, and a corporate board—the Metropolitan Board of Works—to control the main sewers. Boards of both kinds had powers to levy rates, and to seek loans to carry out their work. The Metropolitan Board of Works appointed Sir Joseph Bazalgette as its Chief Engineer, and under the provisions of the Metropolis Local Management Act, he was required, on behalf of the Metropolitan Board of Works:

> . . . to make such sewers and works as they think necessary for preventing all or any part of the sewage within the Metropolis from flowing or passing into the River Thames in or near the Metropolis . . . to be completed on or before 31st December, 1860 and . . . Before the Metropolitan Board of Works commence any sewers or works . . . the plan . . . shall be submitted . . . to the Commissioners of Her Majesty's Works and Public Buildings.

Sir Joseph Bazalgette proposed a scheme of intercepting sewers (figure 22) running parallel to the river and terminating at Barking (Beckton) on the north side, and at Erith (Crossness) on the south side of the river. These would intercept the flows from the existing sewers, but instead of allowing them to discharge directly into the River Thames in London, their flows would be conveyed to the outfalls mentioned, and discharged into that part of the river where it was anticipated that offensive conditions would not be created. He proposed three intercepting sewers on the north and two on the south side (161 km in total).

Sir Benjamin Hall appointed Captain Burstall to examine the plans, but he rejected them on the basis that the discharges were to be made within the metropolitan boundaries, and that this was in breach of the Metropolis Local Management Act. In reply, the board intimated that it had considered plans for siting the outfalls below Gravesend, and that as this would cost £1–2

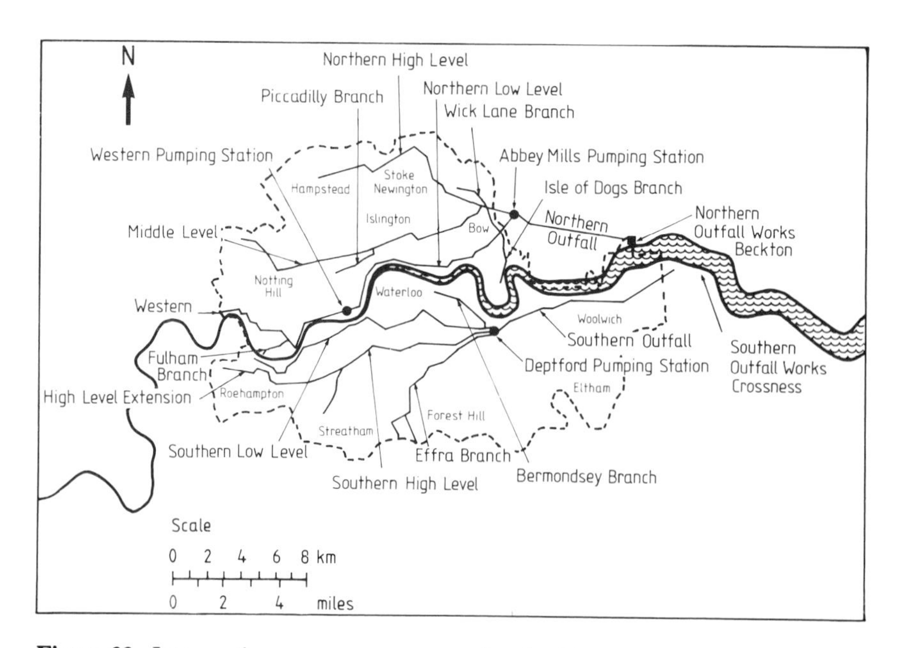

Figure 22. Intercepting sewers constructed by the Metropolitan Board of Works, 1858.

million more and would not benefit ratepayers, it would consider the plan only if the government met the additional cost from national revenue.

The First Commission thereupon submitted the matter to referees, asking them to advise on suitable sites, and also on the possible use of sewage as a source of profit. The appointed referees were Douglas Galton, James Simpson and Thomas E Blackwell. They considered that the sites selected by Bazalgette were unsatisfactory, in that they would not prevent sewage from returning to within the metropolitan boundary, and considered that the nearest sites to London which would fulfil this requirement were the Mucking Lighthouse in Sea Reach for the northern outfall, and Higham Creek in Lower Hope for the southern outfall (figure 20) and estimated that the works would cost £5.4 million. The Board rejected the possible use of sewage for profit, and considered that the only practicable means of disposal was to convey it from highly populated areas as rapidly as possible to be discharged at sea. A prolonged series of exchanges took place, and the Metropolitan Board of Works engaged two consultants, Thomas Hawksley and G P Bidder, to associate with Sir Joseph Bazalgette and to report and evaluate the advantages of the opposing schemes. The situation had reached a deadlock until nature took a hand in 1850s, culminating in the 'Year of the Great Stink'.

As the debates continued, the river became increasingly obnoxious. Benjamin Disraeli (then Chancellor of the Exchequer) introduced in 1858 what became 'An Act to alter and amend the Metropolis Local Management Act 1855, and to extend the powers of the Metropolitan Board of Works for the purification of the Thames and the main drainage of the Metropolis.' The provisions of the former Act were modified as follows:

> The Metropolitan Board shall cause to be commenced, as soon as may be after the passing of the Act and to be carried out and completed with all convenient speed, according to such plan as to them may seem proper, the necessary sewers and works for the improvement of the main drainage of the metropolis and for preventing, as far as may be practicable, the sewage of the metropolis from passing into the river Thames within the metropolis.

The Act became law on 2 August 1858, the 'Year of the Great Stink', and the coincidence gives cause for speculation!

London's First Main Sewerage Scheme

Having been given virtually a free hand by central government, the Metropolitan Board of Works proceeded speedily with the work of coordinating the main sewerage scheme (figure 22), based on the statistics presented in table 4. Sir Joseph Bazalgette proposed for the south side of the river a low-level sewer from Putney, a low-level branch sewer from Bermondsey, a high-level sewer from Roehampton, and a high-level branch sewer from Norwood, all of which were to meet at Deptford pumping station, where the low-level sewage would be lifted and the whole flow would be conveyed to the southern outfall at Crossness for disposal. On the north side of the river there were to be high-, middle- and low-level sewers with branches. The former two were to meet at Old Ford and flow to Abbey Mills pumping station at West Ham, where the low-level sewage would be lifted and the whole flow passed to the northern outfall at Beckton†. The sewage from Fulham and Kensington was originally treated locally at a sewage works to avoid the cost of its conveyance to Beckton, but in deference to the wishes of the residents, it was taken to the western sewer to be pumped to the low-level sewer at the western pumping station.

Table 4. Data adopted by Sir Joseph Bazalgette as the basis of his sewerage scheme.

	North of the Thames	South of the Thames	Total
Area (acres)	31 896	43 546	75 442
Future population anticipated (millions)	2.3	1.2	3.5
Density per acre	72.1	26.4	45.7 (mean)
Sewage per day, assuming $31\frac{1}{4}$ gallons per head (million gallons)	72	36	108
Rainfall per day (million gallons)	178	108	286
Total sewage and rainfall (million gallons)	250	144	394

In order to allow for the diurnal variation in sewage flow (p 160), the sewers were constructed to cater for twice the average flow of sewage. The provision for rainfall was less generous, and to deal with heavy storm flow Bazalgette adopted the idea of using the old sewer outlets to the river, and allowing the sewage to flow over weir chambers into them (figure 23), based on the assumption that the pollution load of sewage would be mitigated by dilution with relatively clean rainwater.

At the outfall works, reservoirs were constructed with capacities to retain six hours' flow of sewage, the one at Beckton extending over 3.8 hectares and that at Crossness over 2.6 hectares. No treatment was given to the sewage, which was then discharged on the ebb tide. It had been shown by experiments with floats that in terms of transporting sewage to sea, limiting the discharge to these tidal conditions was equivalent to siting the outfalls 19 km further seaward if the discharge there were allowed throughout the

† The area adjacent to the northern outfall at Barking was only later known as Beckton.

tidal cycle, although with their greater knowledge of the Tideway, not many modern scientists would agree with this. The northern outfall works and Deptford pumping station were completed in 1864, and the southern outfall works and Crossness pumping station were opened by the Prince of Wales (later King Edward VII) in 1865 (figures 24 and 25). The Abbey Mills pumping station was commissioned in 1868, and the Western in 1875.

Figure 23. Weir chamber under Hammersmith Road, on Counter's Creek sewer and low-level sewer no. 2, looking east. Reproduced from Humphreys 1907 *The Main Drainage of London* (London County Council.)

Bazalgette's scheme was the basis of, and comprises a large part of the present drainage system of London. However, it did mean that the entire sewage flow of the metropolis was discharged into a stretch of river only 3 km long between Beckton and Crossness, and even later, when sewage was treated prior to discharge, the pollution load caused a great depletion of oxygen in the river in that region, resulting in the 'sag' characteristic of poor river quality (see figure 37). With hindsight, had the scheme of Galton and Simpson to extend the sewers to Sea Reach been implemented, with all its faults rectified, much of the subsequent pollution of the Thames in the 1950s would not have occurred. Nevertheless, taken in its own period, with the

Figure 24. The opening of the Crossness works by the Prince of Wales (later King Edward VII), in 1865: the prince is starting the beam engines (from the *Illustrated London News*).

Figure 25. The Crossness (southern outfall) works, 1865 (from the *Illustrated London News*).

knowledge then available, Bazalgette's scheme was an outstanding example of Victorian enterprise. He did not claim for himself originality for the work, stating that it would be difficult from the many suggestions to determine who were the first authors of the various proposals. Certainly, it was due to his energy in transforming the ideas into a practical scheme, and in managing the work in such an efficient manner, that it proceeded to become, with only minor exceptions, the system we have today.

The system did not have sufficient capacity to deal satisfactorily with runoff during storms, and it became necessary to build additional pumping stations to pump sewage directly from sewers to the river at such times. In 1878 pumping stations were built south of the river at the outlets of the old rivers Effra and Falcon, and a further station on the north bank at the Isle of Dogs in 1886. Despite this, further storm relief sewers have had to be constructed, and at present there are 25 major outlets, 14 of which are pumped, and 11 discharge by gravity (figure 26).

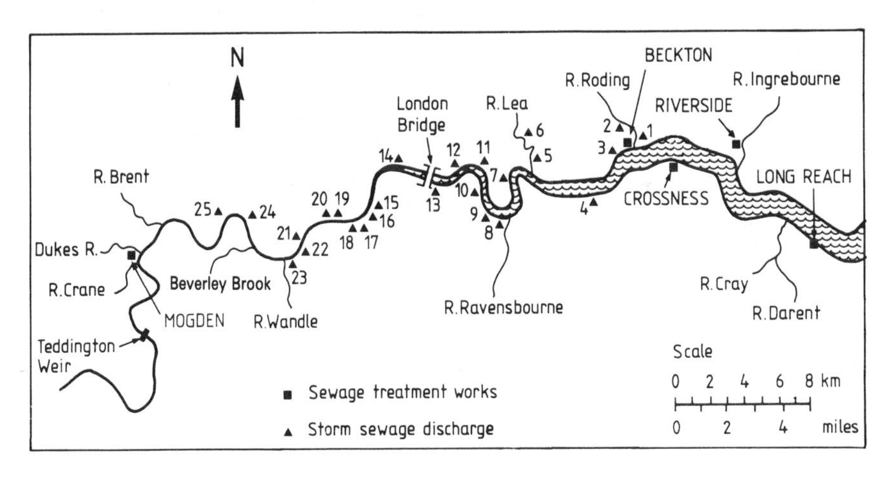

Map ref. no.	Discharge	Method of discharge	Position below London Bridge (km)	Map ref. no.	Discharge	Method of discharge	Position above London Bridge (km)
1	Gascoigne Rd	Pumped	19.0	14	Fleet	Gravity	1.0
2	Folkestone Rd	Pumped	19.0	15	Brixton	Gravity	4.3
3	North Woolwich	Pumped	16.2	16	Clapham	Gravity	4.4
4	Charlton	Gravity	13.1	17	Heathwall	Pumped	5.0
5	Canning Town	Pumped	11.1	18	South-western	Gravity	5.0
6	Abbey Mills	Pumped	11.1	19	Western	Pumped	6.2
7	Isle of Dogs	Pumped	9.9	20	Ranelagh	Gravity	6.4
8	Deptford	Pumped	7.1	21	Counters Creek†	Pumped	8.4
9	Deptford Green	Gravity	6.8	21	Walham Green†	Pumped	8.4
10	Earl	Pumped	5.4	22	Falconbrook	Pumped	9.3
11	Holloway	Gravity	3.6	23	Wandle Valley	Gravity	9.6
12	North-eastern	Gravity	3.0	24	Hammersmith	Pumped	14.9
13	Shad	Pumped	1.1	25	Acton	Gravity	15.7

† These two pumping stations share the same outlet.

Figure 26. Storm discharge outfalls to the tidal Thames.

In addition to the work on the main sewerage system, other plans were made to embank the Thames to improve its flow. Subsequent to an Act of Parliament in 1863 (and further Acts), the work on the embankment of the tidal river was entrusted to the Metropolitan Board of Works, who completed the work under Bazalgette's direction in 1874. This embankment confined the river to a narrower channel and so removed much of the offensiveness caused by the exposure of contaminated mud at low water.

Condition of the River, 1850–1900

Sir Joseph Bazalgette's work on the inception of a sewerage system and the embankment of the Thames in London had temporarily relieved the pollution problems in the metropolis, but before long complaints were being made about pollution and the formation of mudbanks in the vicinity of the outfalls.

The Barking Inquiry

In 1869 the Vicar and other residents of Barking addressed a memorial to the Home Secretary alleging, *inter alia*, that the river constituted a hazard to navigation and to the health of the residents of Barking; and that a bank of sewage mud had been formed in the channel of the river; and that Barking Creek had been closed to vessels exceeding 250 tons.

Sir Robert Rawlinson was therefore appointed by the Home Secretary to hold the 'Barking Inquiry', and after six days he concluded that the allegations were only partly proven, and that no deterioration of health due to the outfalls had been demonstrated. He maintained that the main channel of the Thames had not been reduced in depth nor that Barking Creek had been closed to vessels as described, although he did agree that there had been an accumulation of mud both on the north shore of the Thames and at the mouth of Barking Creek, but the cause had not been proven (Rawlinson 1870).

The Mud-bank Inquiry

In 1878 the Thames Conservancy† alleged that the discharges from the outfalls at both Beckton and Crossness had caused mud banks to form. Under the Thames Navigation Act 1870, the Metropolitan Board of Works was required to keep the Thames free from banks arising from such causes. The matter went to arbitration in 1879, when Sir C A Hartley acted as referee in the case put by Sir Douglas Galton (for the Conservators) and Sir Frederick Bramwell (for the Metropolitan Board of Works). Hartley found that the banks had not obstructed navigation further than by making channels more

† The Thames Conservancy was responsible for management of the Tideway from the time of its institution in 1857 until the formation of the Port of London Authority in 1909.

crooked, and that they had resulted from dredging operations. It was considered that the suspended matter in river water was derived from many sources, and that the sewage discharges contributed only a small amount. It was therefore decided that the Metropolitan Board of Works could not be held responsible.

The Princess Alice *Disaster, 1878*

Another apparently unconnected event was to draw the attention of the public to the condition of the river off the two outfalls (Thurston 1964). It had been a fine day on 3 September 1878, and just as darkness was falling, the 250 ton paddle steamer *Princess Alice* was returning from Gravesend to London with over 800 people on board. As she rounded Tripcock Point on the south bank almost opposite the northern outfall at Beckton, the *Bywell Castle*, a collier of 890 tons, was approaching. The helmsman of the *Princess Alice* was inexperienced; he had only joined the ship that day at Southend on the return journey and had never before handled so large a ship. The paddle steamer was punching the ebb tide, and it was the custom at the time for the vessel to be kept close in to one or other bank, to 'cheat the tide'. The *Princess Alice* rounded the point, keeping to the south bank, but as her navigation lights showed to Captain Harrison of the *Bywell Castle*, the paddle steamer was caught by the tide and swept to the middle of the river. Captain Harrison, considering that she was doing what nine out of ten steamers did, and moving to the protection of the north bank, called for a change of course to starboard, and thereupon struck the *Princess Alice* forward of the starboard paddle box. The *Princess Alice* sank in four minutes, taking with her 650 people who, a few minutes before, had been listening to the ship's band or amusing themselves otherwise. The *Princess Alice* had two life-boats capable of holding only about 50–60 people and there there were only a dozen or so lifebelts. The casualties undoubtedly arose from the fact that at that time most people could not swim and, despite the fact that Henry Belding, mate of the *Bywell Castle*, had actually climbed down ropes from the deck of his ship to that of the paddle steamer, and was urging women to clamber up, they were unable to do so, probably because of the style of their dresses. When the vessel was eventually raised, many bodies were found in the saloon, the doors of which opened inwards and which, in the panic, had been jammed shut by the press of the crowd endeavouring to get out on deck (figures 27 and 28).

An inquest was held (very shortly afterwards because of the lack of refrigeration facilities for the corpses), and during the hearing it was suggested that some of the deaths had been caused or accelerated by the putrid condition of the river. The Medical Officer of Health for the Port of London, Dr Leach, was called, and considered that although the bodies were covered with slime, which they would not have been if the water had been clean, he

Figure 27. The great disaster on the Thames—the collision between the *Princess Alice* and the *Bywell Castle*, near Woolwich in 1878 (from the *Illustrated London News*).

Figure 28. Recovering bodies from the Thames after the *Princess Alice* disaster (from the *Illustrated London News*).

did not think that sewage would poison a man immersed in it for two or three minutes, and in any case men and boys swam in the river, and he had never heard of any ill-effects. It was also considered that the sinking of a heavy body such as the *Princess Alice* would have released much hydrogen sulphide from the river bed, and Dr Leach considered that if a man swallowed much water containing hydrogen sulphide, it would make him vomit. Nevertheless, in reply to a question he said: 'I think I could swim and vomit at the same time, but I have not tried it.'

At the end of the inquest on William Beachey (the first body recovered), the Coroner, Charles Joseph Carttar, attributed the cause of death to drowning as a result of the collision, and recommended the laying down of stringent rules and regulations for navigation on the Thames. The Court of the Admiralty wavered in allocating responsibility (Captain Grinstead of the *Princess Alice* had died bravely on his ship), and eventually both ships were held to be in the wrong. One advantage that came of the tragedy was the introduction of safety precautions and a safe 'rule of the road' for traffic using the Thames.

Thus although the condition of the river had been excluded as the cause of death, the nationwide interest in the disaster drew attention to the pollution, and probably added weight to complaints from other sources.

Improvements at the Outfalls

Probably as a result of local complaints and national concern brought about by the loss of life on the *Princess Alice*, a Royal Commission on Metropolitan Sewage Disposal under Lord Bramwell was appointed in 1882:

> . . . to inquire into and report upon the system under which sewage is discharged into the Thames by the Metropolitan Board of Works, whether any evil effects result thereupon, and in that case what measures can be applied for remedying or preventing the same.

During the hearing, the Commissioners examined 126 witnesses, many of whom were well known experts in their own fields, who were able to inspect the river conditions themselves; 20 000 questions and answers were printed in the minutes. Evidence of a 'popular' nature was collected from police officers, pilots and other river users, which alleged that the river water and mud gave off foul odours, causing nausea and headaches, and that fish had disappeared from the polluted reaches. The Commissioners appeared to attach greater weight to this than to the scientific evidence.

This was probably the first hearing at which scientific evidence was put forward (in this case) by W J Dibdin, Chemist to the Metropolitan Board of Works (later the London County Council) who started a study of the river. Evidence was presented on the levels of polluting substances, such as suspended matter and albuminoid ammonia content. The oxygen demand had been measured in terms of the permanganate value, and a method was available for measuring the dissolved oxygen concentration. The part played by oxygen in destroying organic matter was recognised, as was the lack of dissolved oxygen on the disappearance of fish. It was also accepted that the determination of the movement of sewage upstream by tidal oscillation, as measured by floats, did not entirely represent the facts. Just as the saline sea water could travel upstream by mixing with the fresher water, so could the dissolved matter of sewage discharges.

In the Royal Commission's first report of 1884, the Commissioners found that the discharge of metropolitan sewage did not seem to have any serious prejudicial effect on health, but that in hot, dry weather it caused serious nuisance and inconvenience. Fish had disappeared over a distance of 25 km below the outfalls, and for a considerable distance above them, and there was evidence that wells adjacent to the river had been contaminated. There was no evidence of disadvantageous effects to navigation, but the discharges had added to the detritus in the river. The Commission considered that these ill-effects were likely to increase as long as the contributing population of London continued to grow.

The Commissioners sat again, and produced a second and final report in 1884. They considered the various methods for treating and disposing of sewage including: (i) broad irrigation; (ii) filtration over land; (iii) chemical precipitation, and (iv) chemical precipitation followed by filtration on land. Consideration was also given to moving the outfalls further seaward, but eventually the Commissioners concluded that

(1) it was neither necessary nor justified to discharge crude sewage into the Thames;

(2) at the existing outfalls a precipitation process should be installed immediately to separate the solids from the liquid portion of sewage, and to discharge the latter on the ebb tide, the solids (sludge) being burned, applied to land, or dumped at sea;

(3) the discharge, without further treatment, of settled (precipitated) sewage at the outfalls was not suitable in the long term; it was recommended that it should be applied to land by intermittent, rather than broad irrigation, and if sufficient land could not be found for the purpose, the discharge should be made further downriver; and that

(4) in any future sewerage scheme rainwater should be separated from sewage (i.e. it should be on the 'separate' system).

The Chemist to the Metropolitan Board of Works, W J Dibdin, had experimented in 1884 with chemical precipitation of sewage, and in 1885 it was decided to use lime and protosulphate of iron for precipitation at the outfalls (Dibdin 1887). Precipitation works were therefore built at Beckton (1887–89) and at Crossness (1888–91); plans of the two works are shown in figures 29 and 30. The chemicals arrived by river in barges, from which they were transported to the appropriate part of the works by a small train driven by a steam engine. The works themselves were small communities, the Superintendent and workmen living there (as they did at Crossness almost until 1960). Their houses can be seen in the works plans. The Superintendent had considerable power. At Crossness the school was used during the week for scholastic purposes, but on Sundays it became a place for religious observance; any workman not on a shift, who failed to attend the services was required to account to the Superintendent for his absence!

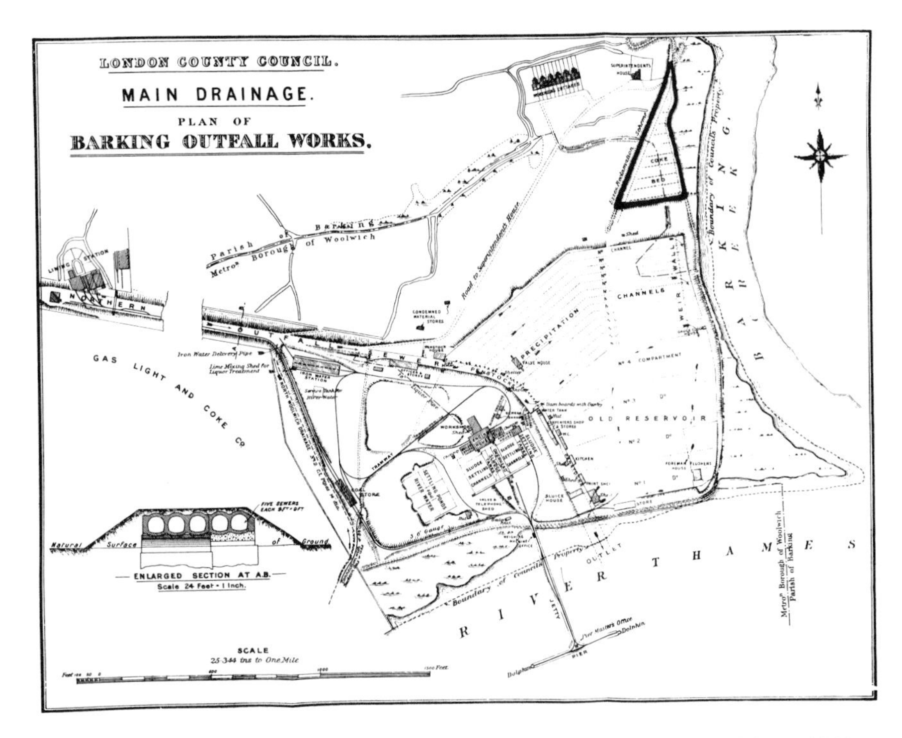

Figure 29. Main drainage—plan of Barking outfall works, 1887–89. The bold line indicates the area covered by Dibdin's one-acre coke bed.

Review of London's Main Drainage

In 1889 the Metropolitan Board of Works was superseded by the London County Council, and on 17 December 1889 it was recognised that further sewers were required. Their Main Drainage Committee was instructed to secure the services of an eminent civil engineer to collaborate with the engineer of the council, Mr (later Sir) Alexander Binnie, to re-examine the whole sewerage system, and to report on the approximate cost of conveying

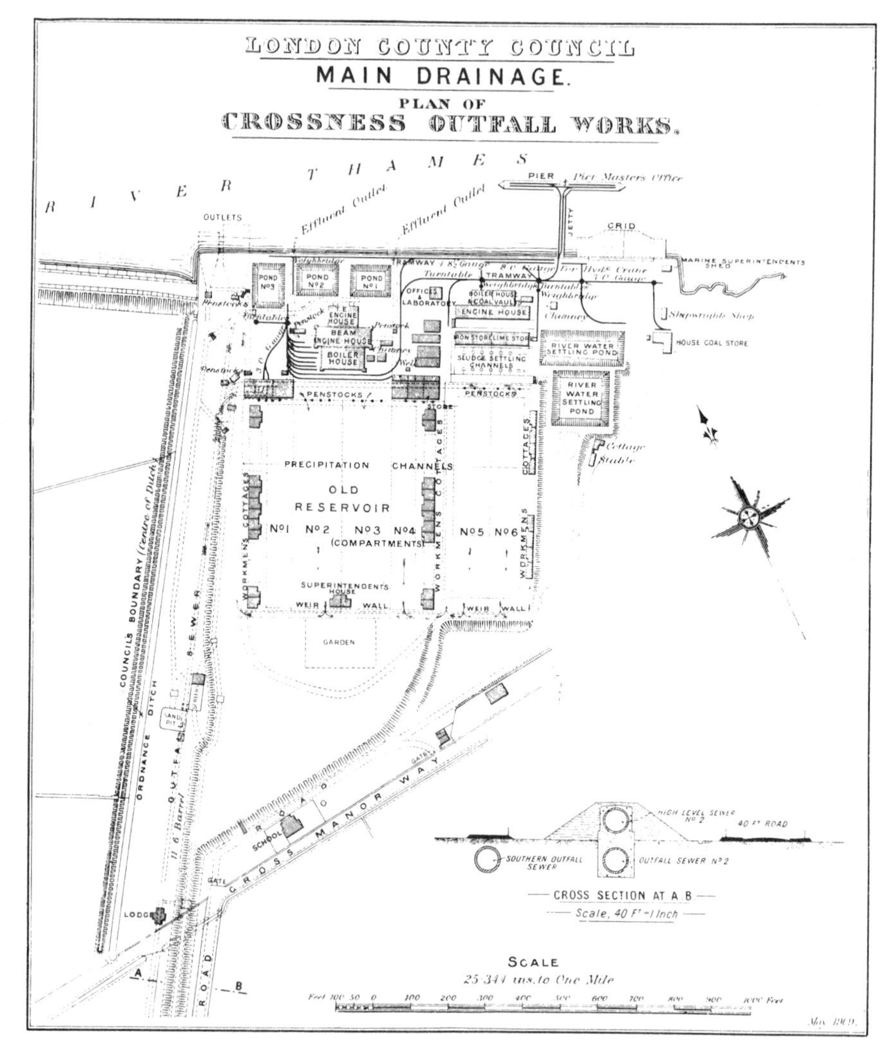

Figure 30. Main drainage—plan of Crossness outfall works, 1888–91.

the whole of the sewage to sea, as recommended in the Royal Commission Report of 1884. Mr (later Sir) Benjamin Baker joined Alexander Binnie, and they reported in February 1891. They considered it impractical to deal with the volumes of sewage arising during heavy storms (Baker and Binnie 1891), other than by letting it discharge to the river. They therefore recommended that new intercepting sewers should be built to prevent flooding of low-lying districts, and to avoid the frequent discharge of only slightly diluted sewage to the river. They did not consider the removal of outfalls to locations further down river to be an urgent matter, and recommended the council to await the results of completing the precipitation works. They expressed some concern, however, about continuing to discharge to the river the whole outputs of Beckton and Crossness (180 mgd† in 1891), when due to increased population the flow had reached 280 mgd (1.27×10^6 m^3/day), which it did in 1910. Little of the new sewerage work suggested by Binnie and Baker had been completed when in December 1897 the Main Drainage Committee called for a further report (Binnie 1899). Binnie again warned of the dangers of not completing the work on new intercepting sewers. Work thereupon progressed slowly owing to the problems with a greatly enlarged city, where considerable care was necessary to prevent subsidence, and to avoid interference with the underground railway system. In 1903 extensive flooding was caused by several periods of heavy rainfall, and the then chief engineer (later Sir) Maurice Fitzmaurice reported, recommending new sewers and pumping stations. By the time of the First World War (1914–18), practically the whole of the scheme of Binnie and Baker had been carried out, together with a large increase in the capacity of Abbey Mills pumping station.

Sludge from the precipitation process was disposed of to sea in the spoil ground at the Barrow Deep, 111 km below London Bridge (see Appendix II).

Early Biological Treatment of Sewage

Although the Metropolitan Board of Works had concluded that in the absence of sufficient land for sewage treatment, chemical precipitation would suffice, Dibdin (figure 31) pioneered experimental work between 1892 and 1905 on filter bed treatment of sewage (Regan 1951; see Appendix II). A one-acre coke bed (figure 29) was used as a contact bed, and smaller filters with different media. Dibdin filled the beds with sewage (using both raw and settled), and allowed them to stand full for an hour, after which they were emptied. When raw sewage was used, although satisfactory purification was achieved, the beds soon became choked, so that the process was considered viable only for settled sewage. Clowes and Houston summarised the work, stating that the process removed 50–80% of the dissolved oxidisable and putrescible matter, compared with only 17% achieved by chemical precipitation (Clowes and Houston 1904). Despite a plan for the installation of large-scale contact beds at the turn of the century, the London County Council made no direct use of this form of treatment.

Condition of the River at the End of the Nineteenth Century

In their report of 1891, Binnie and Baker had stated that there was a marked improvement of the foreshore at the Barking outfall and 'under average

† mgd = millions of gallons per day; 10^6 m^3/day = millions of cubic metres per day.

conditions there is little to reasonably complain of in the state of the river, but . . . at certain times, such as during dry weather and in particular places, the stream is still apt to become very discoloured, and occasionally to emit offensive odours.' Clearly, although the precipitation plant had improved the effluent quality from the outfall works, the lack of sewerage capacity to deal with storms already referred to was causing pollution upstream.

Figure 31. William Joseph Dibdin.

The average composition of the river water off the southern outfall at Crossness from 1885–95 during the summer (third) quarter, when the river was usually at its worst, is given in Table 5 (Dibdin 1896). The reduction in suspended solids over the period is shown in the table, and reflects the lower concentration in the effluent after the precipitation works were installed, confirming that the foreshore would have been cleaner. Removal of these solids would, however, reduce the permanganate value (PV) and the biochemical oxygen demand (BOD) of the effluents discharged, these being measures of the concentrations of pollutants remaining to be oxidised in the river. The PV and BOD values, as shown in the table, gradually increased over the period 1885–95, and this can no doubt be explained by the large increase in sewage flow to these works (figure 32) so that, although the quality of

Table 5. Average composition of river water off the southern outfall at Crossness, 1885–95 (Dibdin 1896). All results in mg/l except dissolved oxygen.

Year	Suspended solids	Cl	PV† (4 hrs *N*/80)	BOD‡	Ammonia		Dissolved oxygen (% saturation)	Temperature (°C)
					Free	Organic		
1885	140	7474	3.2	3.6	1.4	0.84	18.5	18.0
1886	120	4701	2.9	3.3	1.2	0.58	22.4	17.5
1887	79	6430	4.9	5.7	3.2	1.19	15.7	18.7
1888	166	4715	4.4	5.0	2.6	0.87	23.0	16.6
1889	91	4759	4.0	4.6	2.4	0.80	32.3	17.6
1890	36	6473	4.8	5.5	3.2	0.96	33.5	16.5
1891	40	7131	5.1	5.9	3.7	1.17	38.1	16.8
1892	36	6659	4.4	5.1	2.5	1.05	48.2	17.2
1893	77	7531	4.9	5.6	1.6	0.98	36.7	18.5
1894	57	6488	4.7	5.4	1.6	0.84	49.6	16.8
1895	57	6802	5.0	5.7	1.5	0.74	42.5	17.3

† PV: permanganate value.
‡ BOD: calculated as BOD ATU 5 days.

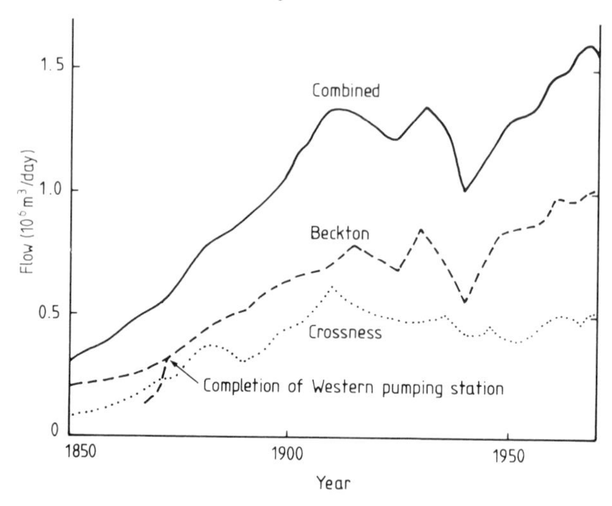

Figure 32. Daily flows of sewage to Beckton and Crossness (flows before 1865 estimated from population data).

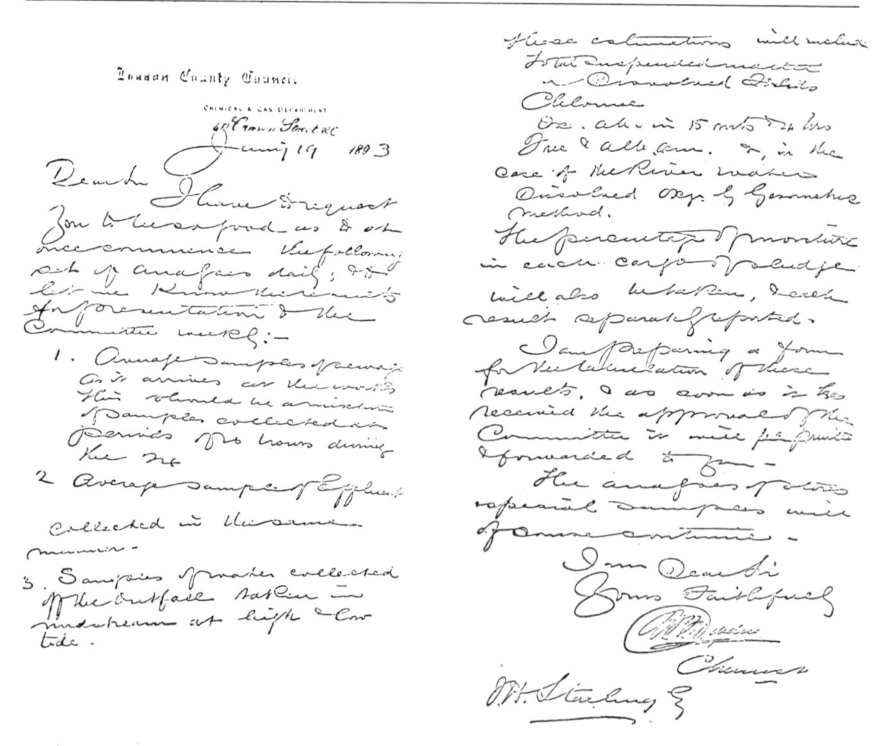
London County Council.
Chemical & Gas Department
Spring Gardens WC
Jany 19 1893

Dear Sir
I have to request you to be so good as to at once commence the following set of analyses daily; and to let me know the results for presentation to the Committee weekly:—

1. Average samples of sewage as it arrives at the works. This should be a mixture of samples collected at periods of two hours during the 24.
2. Average sample of effluent collected in the same manner.
3. Samples of water collected off the outfall station in midstream at high & low tide.

These estimations will include Total suspended matter, Dissolved solids, Chlorine, Ox. abs. in 15 mts & 4 hrs, Free & Alb. Am. &, in the case of the River water, Dissolved Oxy. by Gasometric method.

The percentage of moisture in each cargo of sludge will also be taken, & each result separately reported.

I am preparing a form for the tabulation of these results, & as soon as it has received the approval of the Committee it will be printed & forwarded to you.

The analyses of the special samples will of course continue.

I am Dear Sir
Yours Faithfully
[signature]
Chemist

W. H. Stirling Esq

Figure 33. Letter from Dibdin requesting first regular scientific surveys of the Tideway, 1893, together with a transcription (*right*).

effluent had improved, the load on the river was greater because of the greater volumes discharged. Despite this, the dissolved oxygen concentration was higher which indicates, as Dibdin pointed out, 'the merest surface of the mud on the foreshores and bed of the river at once accounts for the utilisation of [these] hundreds of tons of oxygen', and when this was removed the dissolved oxygen increased. The higher pollution loads held in solution and suspension, and shown after 1887 by the increased PV and BOD of the water, would be removed by dilution and oxidation as they were carried downstream, and would not be as significant as an equivalent load retained in mud on the foreshore.

Until the 1880s, relatively little scientific monitoring of the river had been carried out, but in 1893 Dibdin, then chemist to the London County Council, became responsible for the introduction of regular monitoring of the whole of the tidal river (figure 33). He carried out a detailed examination between 1893 and 1894, and related his chemical analyses for suspended solids, permanganate value, dissolved oxygen content and chloride, with the volume of a relatively unpolluted upland flow of the river above Teddington weir, and the tidal range at London Bridge (Dibdin 1894). He was the first to

---

Dear Sir, January 19th 1893

I have to request you to be so good as to at once commence the following set of analyses daily; and to let me know the results for presentation to the Committee weekly:—

1. Average samples of sewage as it arrives at the works. This should be a mixture of samples collected at periods of 4 hours during the 24.
2. Average samples of Effluent collected in the same manner.
3. Samples of water collected off the Outfalls taken in midstream at high and low tide.

These estimations will include

Total suspended matter and Dissolved Solids

Chlorine

Ox[ygen] ab[sorbed] in 15 m[inu]t[e]s and 4 hours.

Free and alb[uminoid] Am[monia] and, in the case of the River Water

Dissolved Oxy[ge]n by Gasometric method.

The percentage of moisture in each cargo of sludge will also be taken, and each result separately reported.

I am preparing a form for the tabulation of these results and as soon as it has received the approval of the Committee it will be printed and forwarded to you.

The analyses of stores and special samples will of course continue.

I am Dear Sir
Yours faithfully
W. J. Dibdin
Chemist

J. H. Starling, Esq.

study the uptake of oxygen by polluted river samples, and concluded that more oxygen was taken up by suspended than by dissolved materials. He considered that a flow of 500 mgd (2.27×10^6 m^3/day) over Teddington Weir would be necessary to prevent polluted water passing up the estuary with the tide. When sludge vessels (ships which transport sewage sludge for disposal at sea) were introduced, laboratories could be installed on them; these offered a convenient method of monitoring the river which is still used to this day (see Appendix V). Undoubtedly without the data accumulated since 1893, when Dibdin instigated regular monitoring, the restoration of the tidal Thames in the twentieth century would have been very difficult indeed.

The improved conditions demonstrated chemically by Dibdin were confirmed by the improved aquatic life. After 1890 bleak, dace and roach were plentiful above Putney, whereas previously none had been seen below Kew. Two years later whitebait reappeared at Gravesend, and flounders again populated the river. The levels of dissolved oxygen were as shown in figure 37.

The Second Deterioration, 1900–50

Deterioration of River Quality

As London passed into the twentieth century its population and water demand increased (Figure 34). The tidal river deteriorated in the first half of the century to become a putrid and noisome river in which no aquatic life could survive and which was a nuisance to those using or living near it. Fortunately, Dibdin had introduced regular river sampling in the 1890s, and this has been continued to the present day, thus preserving the record of changing river quality.

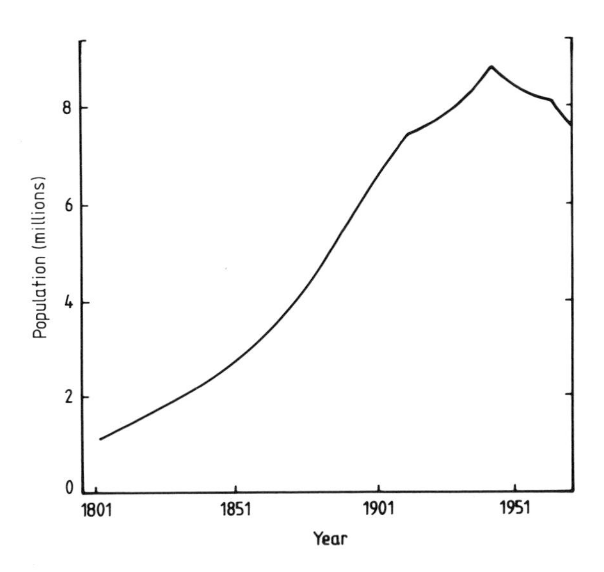

Figure 34. Trends in the population of Greater London.

A rough indication of the deterioration can be seen in figure 35, where the minimum point of the curve for average dissolved oxygen in the third quarter

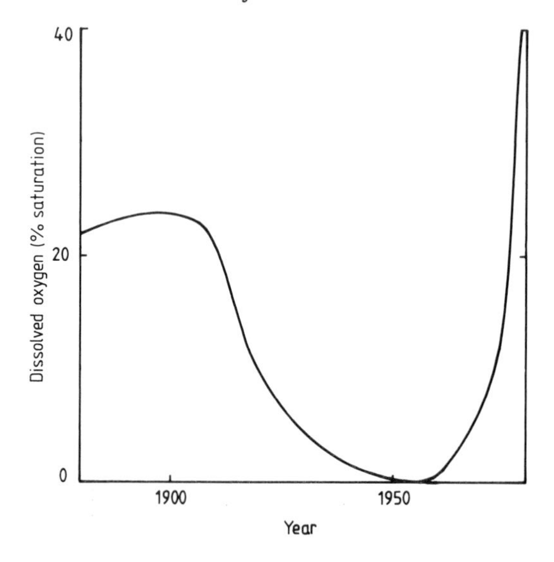

Figure 35. Third quarter, average dissolved oxygen curve, lowest point.

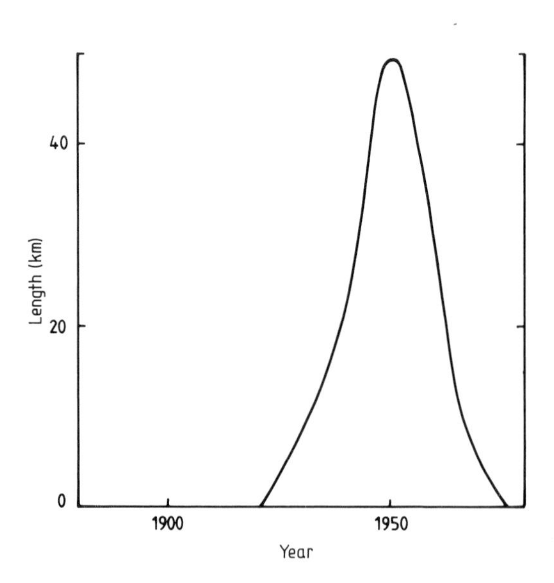

Figure 36. Length of river with dissolved oxygen (percentage saturation) below 5%.

(usually the worst conditions) is plotted for the years 1880–1980. The smooth curve demonstrates that there was a rapid deterioration after 1910, reaching near-anaerobic conditions in 1935 and complete anaerobicity in 1950. Figure 36 is a smoothed curve in which the length of river with dissolved oxygen below 5% saturation in the average third quarter curve is shown on a yearly basis. The significance of choosing this concentration of 5% is that it is known (see Appendix I) that when the concentration of dissolved oxygen falls to this level, the micro-organisms supplement it by using oxygen combined in dissolved salts such as nitrate to continue the river's self-purifying process. Figure 36 presents a similar picture of rapid deterioration after 1920 to achieve the worst conditions in 1950. Figures 35 and 36 give a good indication of river quality at the worst point, but to examine overall quality we must turn to figure 37.

In figure 37 the concentrations of dissolved oxygen along the length of the Tideway are shown, and the 'sags' which occur in the region of the greatest pollution. In comparing these 'sag' curves from year to year, note must be taken of the upland flow from above Teddington Weir: the greater this is, the further the sag will be moved downstream and, because the volume of the estuary increases exponentially seawards (see Appendix V), the shape of the curve will change—the leading edge will tend to rise and the width will diminish. Furthermore, the greater the flow of the clean upland water, the more dissolved oxygen it will make available for the self-purification processes in the Tideway.

There appear to be three distinct phases in the downward trend in water quality, as shown in figure 37. In the first phase (1893–1910) the river had a level of dissolved oxygen of not less than 25% saturation in all places, and therefore would have been capable of supporting a good population of coarse fish. Over the period reviewed, the human population and the resultant polluting load of effluents from the two metropolitan treatment works at Beckton and Crossness increased rapidly (figure 32). That the river maintained its quality was probably due to the fact that there was still sufficient capacity at the works installed in 1889–94 to produce effluents of a quality which was still within the river's self-purifying capacity to absorb, despite their increase in volume.

In the second phase (1915–30), despite a slight increase in sewage flow from the metropolis, the level of dissolved oxygen fell almost to zero within stretches of river some 10 km above and below the major sewage outfalls. It is therefore probable that at some time shortly before 1915 the capacity of the river to accept the effluent loads was exceeded, and it was no longer capable of providing the dilution and dissolved oxygen required. It may also have been due to the decision to discontinue the use of chemical precipitation at Beckton and Crossness at this time, and to rely on unassisted sedimentation, resulting in the production of effluents of poorer quality which increased the load on the river. These worsened river conditions would have affected fish

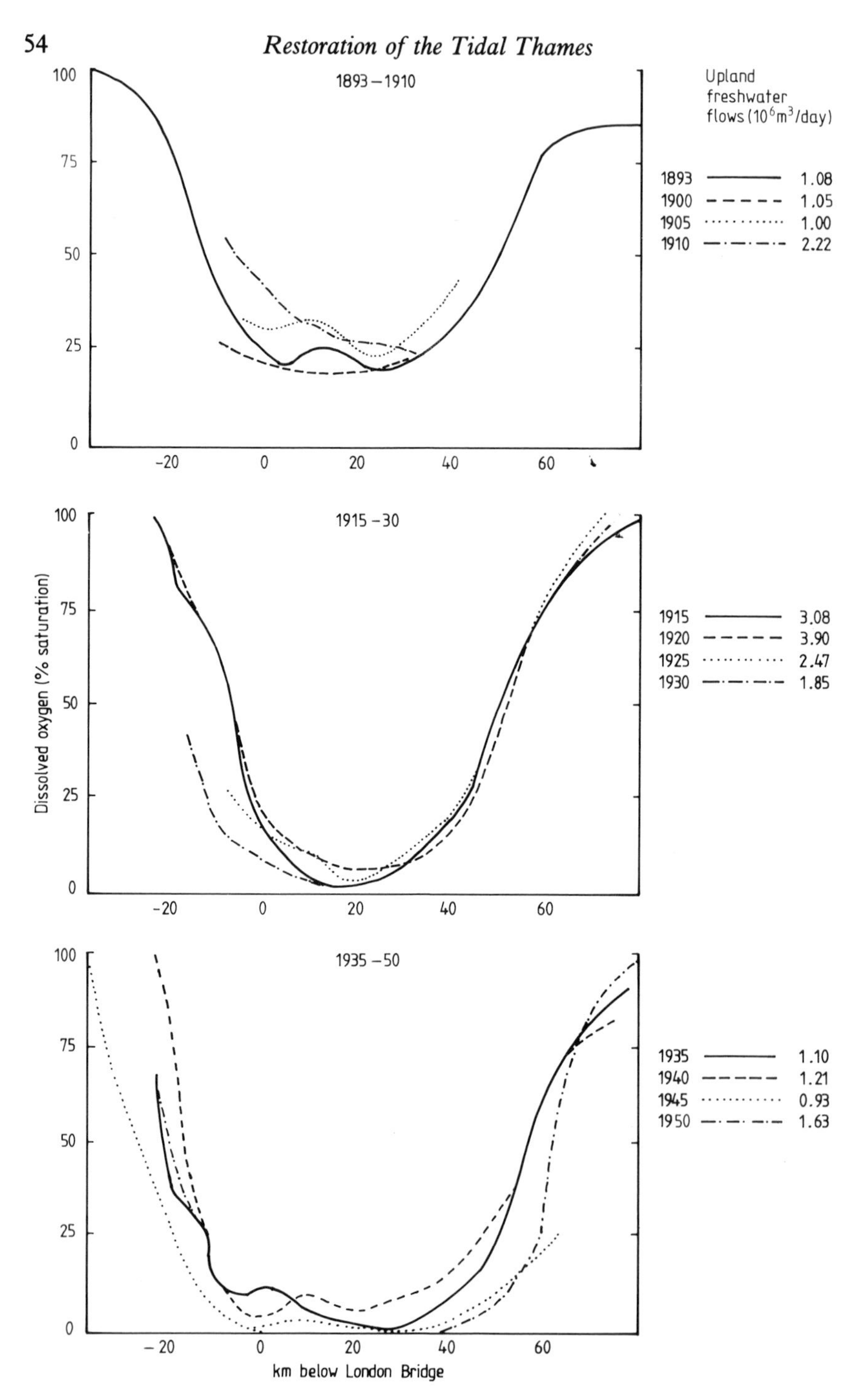

Figure 37. Dissolved oxygen concentrations, 1893–1950, third quarter averages.

populations adversely, and the latter would have been present only sporadically, if at all, in the Thames in London.

During the third phase (1935–50) there was a considerable widening of the lower part of the sag of the dissolved oxygen curve. This undoubtedly resulted from the effects of the spread of urban population, and the consequent discharge of sewage effluents from an increased number of small sewage works which had been developed by that time along a much greater length of river.

Sources of Pollution

With the growth of London from 1900–50, the quantity of polluting matter of all kinds discharged to the Tideway had increased. Such pollution was from the so-called 'non-point' discharges made over an undefined area, such as run-off from agricultural land, as well as (for the most part) from 'point' discharges derived from sewage works outfalls, sewers from industrial premises discharging directly to the river, storm water overflow from surface water sewers (containing rainwater run-off), and storm sewage overflow from sewers of the 'combined' system. These major sources are described below.

Sewage Treatment Works

By the time the tidal river had reached its worst condition in 1950–53, effluents were being discharged from the works listed in table 6. It can be seen that the major pollution loads—with the exception of that from West Kent† (Long Reach) sewage treatment works—result from sewage works in the Greater London area. Moreover, since the volume of the estuary increases almost exponentially with the distance seaward, the dilution afforded to any discharge increases in a similar manner. Hence the reduction in concentration of dissolved oxygen caused by the discharge of 10 tonnes/day of an oxygen-consuming load decreases sharply with the distance seaward (figure 38). The effect on the river of effluents from works in Essex and Kent was therefore very small compared with that from works in Greater London. Long Reach works, being only just downstream of the boundary of Greater London, did contribute a significant pollution load; it was the last major sewage works to be upgraded (in 1978).

The two largest works in the Greater London area are Beckton and Crossness; the quality of their effluents has always had a greater effect than that of any other on the condition of the Tideway. The variation in sewage flows to the two works, the BOD loads arriving in the sewage, and in the effluents discharged, are shown in figures 39 and 40, respectively. The relative

† West Kent sewage works was renamed Long Reach when taken over by the Thames Water Authority in 1974.

Table 6. Sewage works discharging into the Thames Tideway, 1950.

Sewage works	Position below London Bridge (km)	North or south bank	Flow ($10^6 m^3$/day)	BOD (mg/l)	BOD load (tonnes/day)
Greater London					
Ham	−28.8	S	0.00068	60	0.041
Mogden	−25.0	N	0.378	17	6.4
Richmond	−20.2	S	0.026	110	2.9
Acton	−16.3	N	0.014	530	7.4
Beckton	19.0	N	0.90	240	210
East Ham	19.5	N	0.020	120	2.5
Crossness	22.7	S	0.45	180	80
Dagenham	25.2	N	0.040	120	4.7
Essex and Kent					
West Kent	32.3	S	0.103	230	24
Stone	34.8	S	0.0014	115	0.16
Swanscombe	37.0	S	0.0009	200	0.18
Northfleet	41.3	S	0.0027	100	0.27
Tilbury	45.0	N	0.013	100	1.3
Gravesend	46.3	S	0.0055	265	1.5
Stanford-le-Hope	53.5	N	0.0018	26	0.05
Corringham	59.7	N	0.00068	33	0.02
Pitsea	59.7	N	0.00068	33	0.02
Canvey Island	61.8	N	0.00068	360	0.24
South Benfleet	66.7	N	0.0011	26	0.03
Leigh-on-Sea	66.7	N	0.0020	95	0.19
Southend-on-Sea	74.3	N	0.030	260	7.9
Total			1.99		350

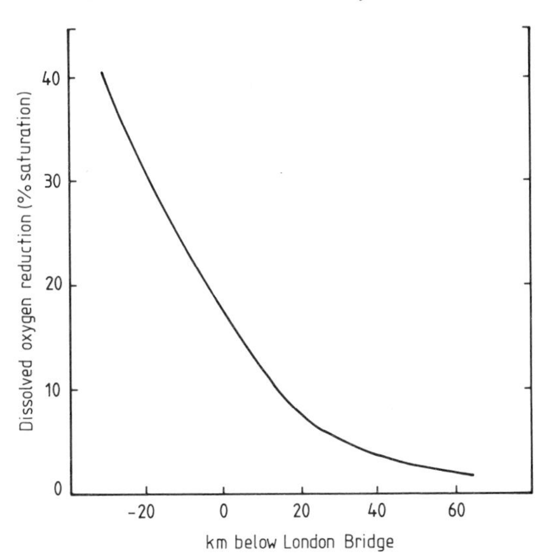

Figure 38. Maximum reduction in dissolved oxygen due to discharge of 10 tonnes BOD at different points along the Tideway.

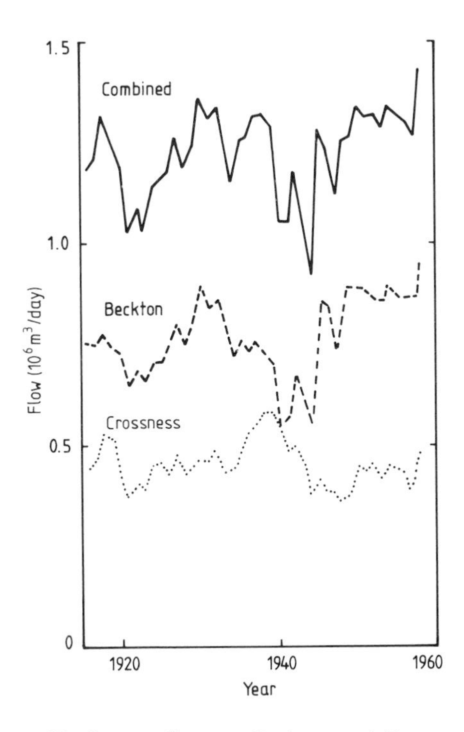

Figure 39. Sewage flows to Beckton and Crossness.

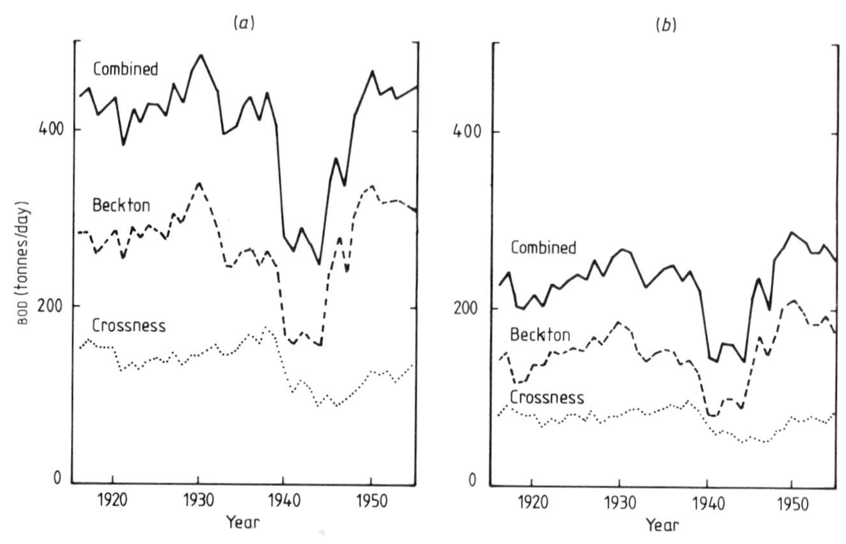

Figure 40. BOD loads (*a*) arriving in sewage, and (*b*) discharged in effluents at Beckton and Crossness.

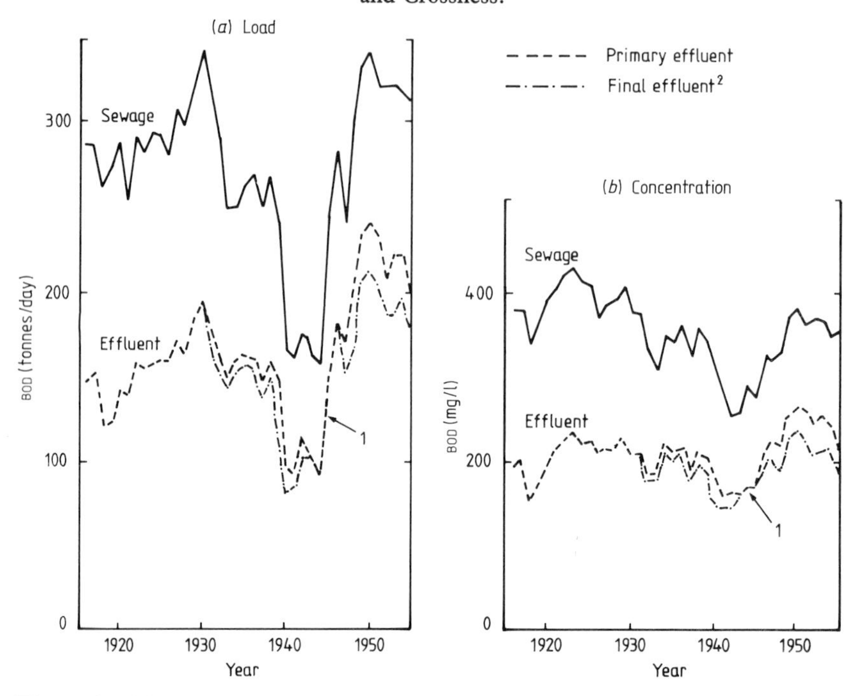

Figure 41. (*a*) Load and (*b*) concentration of BOD in sewage and effluent at Beckton. 1, activated sludge plant was shut down during the war years 1939–45 due to fuel shortages. 2, final effluent is that in which part has been subjected to activated sludge treatment. Where BOD values are not available they have been calculated from PVs.

BOD levels in the sewage and effluent at the works indicate their performance in sewage treatment (figures 41 and 42).

In respect of the loads discharged from the works (figure 40) there were regular increases from the turn of the century until 1930, which probably account for the deterioration in river quality between 1915 and 1930 (figure 37). After this time the load from Beckton decreased, whilst that from Crossness continued to increase slightly. At Beckton the most significant factor was probably the reduction in sewage flow (figure 39) occasioned by the movement of some of the population of central London to the developing suburbs. Such a movement did not occur south of the river, perhaps due to differences in lifestyles, and to the lower industrial development. Not only did the flow of sewage decrease at Beckton and increase at Crossness, but also the strength of the Beckton sewage decreased (figure 41), while that at Crossness remained almost constant (figure 42). Thus the pollution load (a product of flow and strength) of sewage at Beckton fell even more sharply, and at Crossness increased steadily (figure 40).

The treatment provided when both works were constructed in the 1880s involved the precipitation of sewage during sedimentation by the use of chemical coagulants (lime and ferrous/ferric salts), but this practice was abandoned after 1915. Crossness continued to rely on unassisted precipitation of sewage by sedimentation until 1963. At Beckton, however, a secondary treatment plant was built between 1932 and 1938 which comprised a paddle-aeration activated sludge plant designed to treat about a third of the flow. The plant was different from modern systems in that the activated sludge was aerated for about $2\frac{1}{2}$ hours and then mixed with settled sewage and circulated through a rather complicated system of two-decker channels for about $2\frac{1}{4}$ hours. No aeration other than that obtained from the surface

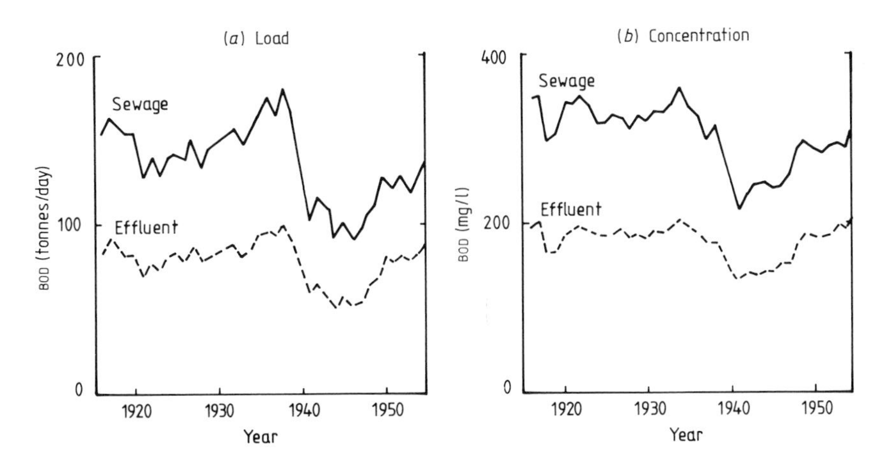

Figure 42. (*a*) Load and (*b*) concentration of BOD in sewage and effluent at Crossness.

was available. The activated sludge was then separated in final settlement tanks and returned to the aeration channels. Two of the six-paddle channels were in operation in 1932 but as a result of bomb damage, and for other reasons, the remainder only came into operation in 1946 after a period of three years (1943–46) in which the whole activated sludge plant had been shut down because of fuel shortages. The effect of this plant can be seen in figures 41 and 43; it removed only about 10% of the Beckton sewage load, and 6% of the combined loads from the two major works at its best performance, and had little value in improving the quality of the Tideway.

The flow and strength of sewage at both works fell during the Second World War (1939–45); this in itself should have resulted in an improved river condition but the effect was obviously obscured by the effect of other discharges, including sludge dumping at Mucking Flats (p 62), and the results of bomb damage to sewers and sewage treatment plants throughout London.

In the post-war period the sewage appears to have been more difficult to treat than that of before the war, possibly due to the widespread use of synthetic detergents. At a sewage works their effect would be to retain in suspension in the effluent substances which otherwise would have been

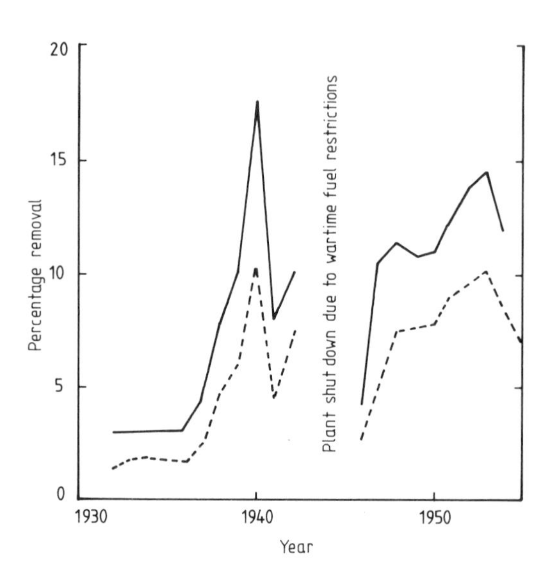

Figure 43. Performance of the first activated sludge plant at Beckton, 1932–50.

$$\text{Full curve} = \frac{\text{BOD load removed} \times 100}{\text{BOD load in primary effluent}} \text{(tonnes/day).}$$

$$\text{Broken curve} = \frac{\text{BOD load removed}}{\text{BOD load in sewage}} \text{(tonnes/day).}$$

precipitated during sedimentation. At Beckton, where there was a diffused air-activated sludge plant in operation, non-biodegradable surface-active agents such as those described on p 69 would have reduced the rate of oxygen transfer, and the plant could be expected to operate at only 70% efficiency.

By 1950 (figure 40*b*) the total pollution load discharged from Beckton and Crossness was about 10% greater than that of the period immediately after the war. The river off the outfalls deteriorated very sharply (figure 44), partly due to this, but also as a result of the effects of synthetic detergents on the self-purification processes of the river (Appendix I). The disadvantage of siting the two metropolitan outfalls so closely together that they discharged into a stretch of river only 3.2 km long was now apparent, in that there was insufficient river water to effect satisfactory dilution and to remove the pollution by self-purification. In fairness to the designer (Sir Joseph Bazalgette) however, he had foreseen this and had proposed the transfer of at least part of the effluent downstream, when its volume had increased sufficiently. The deterioration in river quality from 1930–50, much of which was due to the greater pollution loads from Beckton, can also be seen in figure 37.

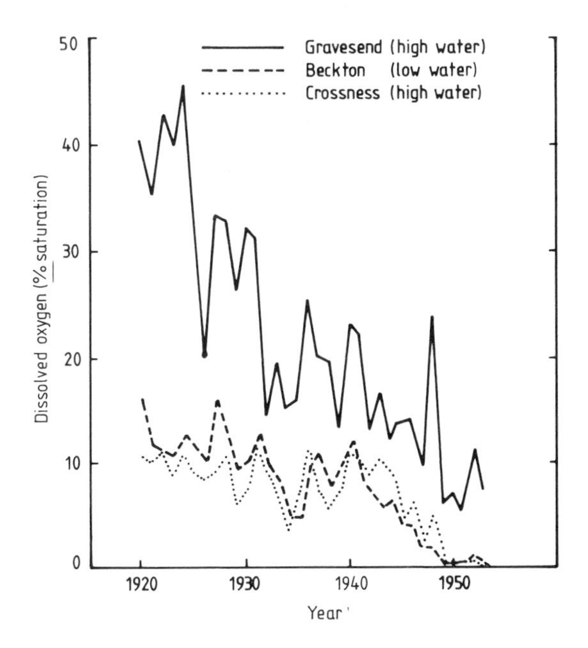

Figure 44. Dissolved oxygen (% saturation) at Gravesend, Beckton and Crossness (third quarter averages).

On 29 October 1940 the sludge vessel SS *G.W. Humphreys* struck a mine in the Thames estuary and sank. In an attempt to protect other similar vessels the Admiralty therefore ordered that sewage sludge should be dumped further upstream. Sludge had previously been discharged first in the Barrow Deep and later in the Black Deep (both about 110 km below London Bridge), but after the *G.W. Humphreys* incident it was to be discharged at Mucking Flats, about 57 km further inland. The site chosen was below Gravesend, and therefore in that part of the river where the volume increased rapidly and so afforded reasonable dilution. However, the daily discharge still represented a highly polluting load that was approximately equal to that of the combined effluents of Beckton and Crossness. It is difficult to determine from the records what effect this discharge had on the river, since dumping arrangements interfered with monitoring, which was also carried out from the sludge vessels, but the effect can be assessed from the use of the mathematical model.

Figure 45 shows the extent of the reduction in dissolved oxygen which the model predicts for third quarter minimum flow conditions. When the observed third quarter curve for 1940 is adjusted, by subtracting the oxygen requirements of sludge dumping (figure 46), the curve is similar to that actually found for 1945, taking into account the slightly reduced upland flow in the latter year. It therefore seems that this operation could account for much of the difference between the curves for 1940 and 1945, and could also explain why the reduced load discharged from Beckton and Crossness did not effect an improvement in river quality, although some improvement at the outfalls would have been expected. With the return of peacetime conditions in 1945, the combined flow from the two works increased to pre-war

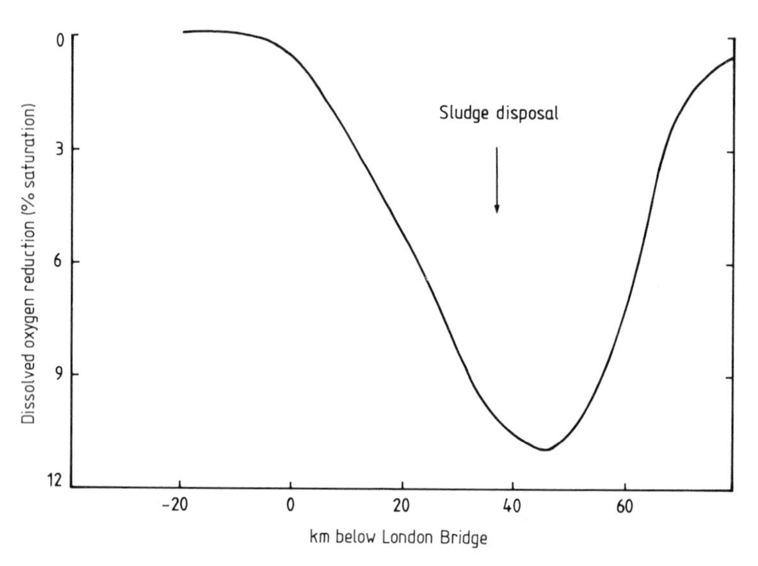

Figure 45. Depletion of dissolved oxygen caused by sludge disposal at Mucking Flats, 1940–45.

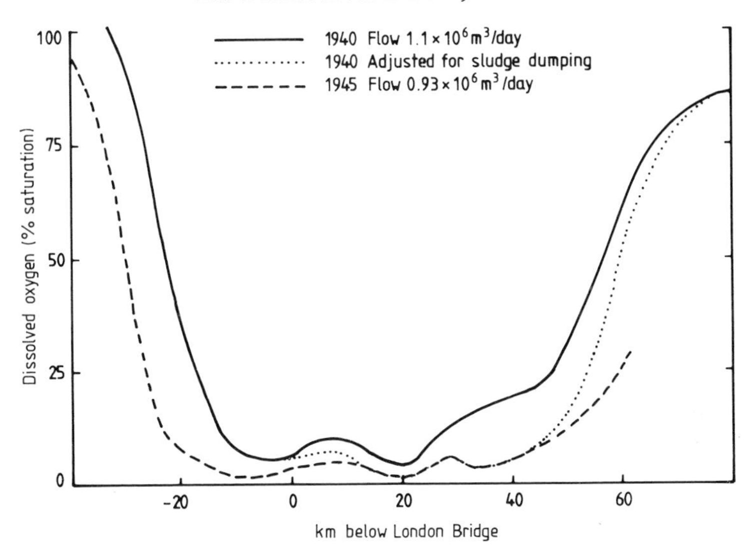

Figure 46. Dissolved oxygen sag curves, 1940 and 1945 (third quarter).

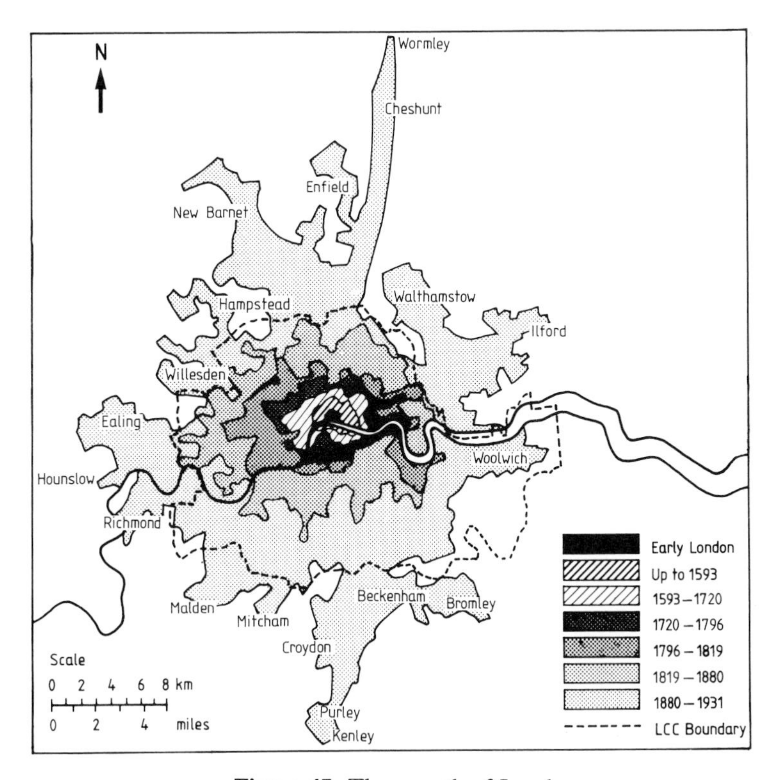

Figure 47. The growth of London.

levels and the load to about 10% above the pre-war load. This was just one factor in the decline of the river in the 1950s.

Between 1880 and 1930 the population of London increased rapidly, and as a result the city expanded beyond the county boundary (figure 47) and more sewage works had to be built which discharged into either the tidal Thames or its tributaries within the metropolitan area. The area of Greater London outside the county boundary of 1931 occupied 4580 km^2 within a circle of approximately 40 km radius centred on Charing Cross. Within this area there were 182 sewage works; on average one for every 16.2 km^2 (table 7). Thus, in addition to the 108 works which discharged directly (or indirectly by way of tributaries) into the tidal Thames, there were 74 other works discharging to tributaries of the Upper Thames within 20 km of Teddington Weir. It has already been indicated (figure 38) that the discharge of polluting loads into the upper part of the tidal river could have very significant effects. Possibly because of the sensitivity of the river in this region, but more probably because of the need for more efficient management, legislation was introduced early in the century, as a result of which the quality of the upper Tideway was not allowed to deteriorate as it had done off the metropolitan outfalls at Beckton and Crossness.

Table 7. Sewage works within 40 km of Charing Cross, 1930.

Area	River catchment	Area (km^2)	Number of sewage disposal works	
			Total	Area per works (km^2)
1	East of Roding	230	16	14.4
2	River Roding	208	17	12.2
3	River Lee	487	37	13.2
4	Rivers Brent and Crane	195	23	8.5
5	River Colne†	589	39	15.1
6	River Wey†	314	14	22.4
7	Rivers Mole† and Hogsmill†	332	21	15.8
8	Beverley Brook and Rivers Wandle, Ravensbourne, etc.	593	15	39.5
	Total	2948	182	Average 16.2

† Tributaries of the Upper Thames.

Storm Sewage Overflows

It would not be an economic proposition, even were it possible, to build a sewage treatment works to treat in times of storm the extremely large volumes of liquid received into sewers operating on the combined system, i.e. receiving both domestic sewage and drainage of storm water. If works were so designed, a large part of the plant would remain unused for most of the time, and there would be an uneconomic use of capital. It is usually the practice to design a works to treat three times the dry weather flow (DWF), and to store flows above this (up to six times the DWF) for future treatment; flows above six times DWF are discharged directly to watercourses. The Thames has many storm sewage discharge outlets (figure 26), some of which discharge by pumping to the river, while others receiving flows from higher ground discharge by gravity. Even now, it is not possible to measure the flows, or sample the strength of gravity discharges. Their total load to the tidal Thames is now thought (as a very rough estimate) to be 15% of that of adjacent pumped discharges. Storm sewage discharges have always represented a problem, as they introduce a heavy pollution load for a short period, which can have a disastrous effect on the biota of the river; it is only recently that an economical method of combating these discharges has been found (p 134).

The quantities of storm sewage discharged between 1900 and 1950 from pumping stations are shown in figure 48, as daily averages for each year. Whilst this is not a very realistic indication of the magnitude of flows during particular storms, it gives some account of the relative responsibility of storm sewage for the deterioration of the river. It cannot be said that this was a major contribution to the river's decline between 1900 and 1950, but it undoubtedly exacerbated the poor conditions resulting from the highly polluting effluents discharged regularly. The daily flows of storm sewage shown in figure 48 represent only a small fraction of the quantities discharged from Beckton and Crossness.

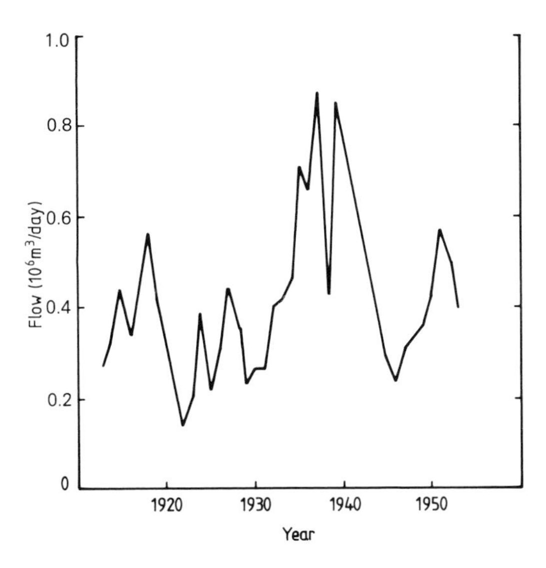

Figure 48. Storm sewage overflows, pumped discharges, 1913–53.

Table 8. Industrial discharges to the Thames Estuary (other than from power stations), 1952 and 1962.

Nature of industry	Position of discharge below London Bridge (km)	North or south of river	BOD load (tonnes/day) 1952	1962
Sugar refinery	−15	N	0.02	0.02
Glucose and maize products	−10	S	0.03	0.07
Gas works	12	S	0.9	0.5
Flour mill	12	N	0.06	nil
Flour mill	12	N	0.09	nil
Sugar refinery	15	S	0.21	0.06
Sugar refinery	15	N	0.06	0.04
Edible-oil refinery	15	N	0.1	0.2
Gas works	18	N	5.5	0.8
Distillery	22	N	25.8	nil
Edible-oil refinery	20	S	0.3	0.25
Chemical works	30	S	0.5	0.6
Paper mill	30	S	0.2	0.7
Paper mill	30	S	0.9	2.7
Chemical works	31	N	0.2	0.4
Board mill	32	N	6.2	6.1
Margarine factory	33	N	2.0	3.0
Paper mill	36	S	0.8	1.9
Soap works	37	N	0.2	0.3
Paper mill	41	S	2.2	2.2
Paper mill	42	S	0.8	1.2
Paper mill	43	S	0.9	1.4
Petroleum depot	55	N	0.5	0.5
Petroleum products	55	N	Not known	8
Petroleum products	58	N	Nil	4
Petroleum products	59	N	Not known	1.6
Total			48	36.5

Industrial Effluent Discharges

The industrial effluents discharged to the Thames Tideway have never been as polluting nor have they caused such intractable problems as those made to rivers in the north of England. In 1952 and 1962 the main industrial discharges (other than those from power stations) were as shown in table 8, and amounted to only 9% of the total pollution load over this period.

Table 9. Major sources of pollution entering the Tideway, 1950.

Source	Position below London Bridge (km)	Load (tonnes/day)			Percentage of total BOD load
		BOD	Organic nitrogen	Ammonia nitrogen	
Mogden	−25.0	6.4	0.8	0.9	1.8
Richmond	−20.2	2.9	0.1	0.3	0.7
Acton	−16.3	7.4	0.6	0.8	2.0
Beckton	+19.0	206	8.8	22.6	55.4
Crossness	+22.7	79.9	4.0	9.3	21.5
Long Reach	+32.3	23.7	1.0	2.6	6.3
Small works (total)		12.6	0.5	1.3	3.4
Industrial (total)		32.7	3.5	—	8.8

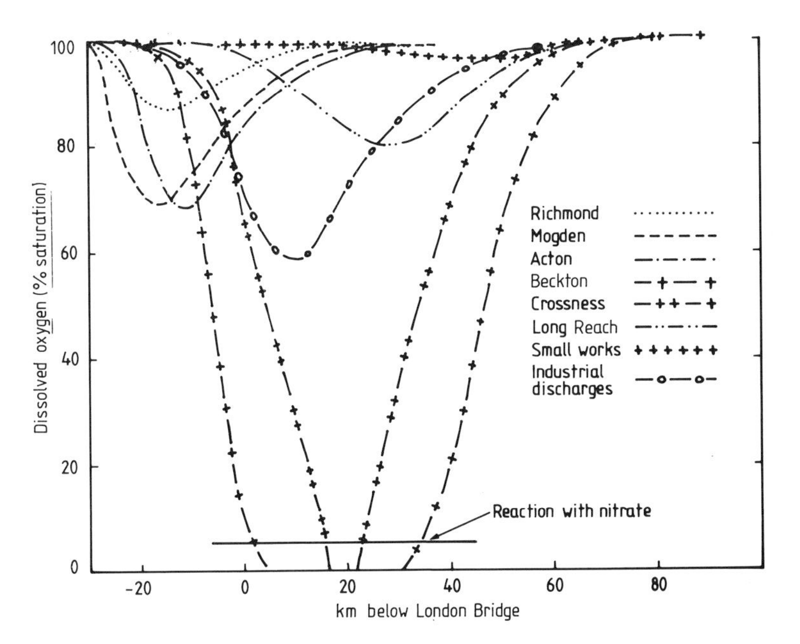

Figure 49. Effect of major sources of pollution of the Tideway—reduction in levels of dissolved oxygen.

The major sources of those polluting discharges which reduced the levels of dissolved oxygen in the river have already been described, and their relative magnitudes in 1950 are given in table 9. The oxygen uptake of these loads and their effects on the depletion of dissolved oxygen of the river can be modelled mathematically, as seen in figure 49. It is clear that Beckton and Crossness were of by far the greatest consequence, but the loads from Mogden and Acton can be seen to need control even though they were comparatively small, because of their positions, for reasons which are apparent if reference is made to figure 38.

Reduction of Dissolved Oxygen

Not only does a river become polluted by the introduction of biodegradable wastes which consume oxygen, but it can also be affected deleteriously by activities which interfere with its supplies of oxygen. This interference, of course, affects the river biota adversely, and also reduces the quantity of biodegradable load which the river can receive if it is to remain viable.

The sources from which the river derives the major part of its dissolved oxygen are set out in table 10. It can be seen that the principal source of oxygen is that obtained from the atmosphere by absorption at the water surface. Other factors being equal, the rate of solution will be proportional to the area of the interface between the water and air. Consequently, this is a source which is very much subject to weather conditions; for example, high

Table 10. Sources of oxygen entering the Thames estuary.

Source	Amount of oxygen			
	(tonnes/day)		(%)	
Upper Thames	70		9	
Tributaries	12		1.6	
Rainfall	2		0.3	
Effluents	3–9		0.8	
From sea	80		10.4	
	—		—	
Total from advective sources		170		22
Absorption from atmosphere†		600		78
Photosynthesis‡				

† The rate of absorption is 24 tonnes/day for 1 mg/l deficiency, for each 1 cm/h of exchange coefficient. On average for the Thames Tideway with a deficit of 5 mg/l, and an exchange coefficient of 5 cm/h, the uptake is 600 tonnes/day.

‡ The input from this source is too variable to quote any average as it can vary as widely as from 1 to 850 tonnes/day.

winds produce waves, which present a large and rapidly changing surface to the atmosphere. On average, about 78% of the oxygen of the Tideway is obtained from surface aeration. The rate of solution of oxygen at the surface can, however, be greatly retarded by the presence of surface-active agents such as synthetic detergents (see Appendix I).

Synthetic Detergents

Before 1939 there was very little use of synthetic detergents; those which were used were largely biodegradable, containing mainly alkyl sulphates as the active material. After the war, however, there was a rapid increase in the use of synthetic detergents (table 11) containing alkyl aryl sulphonates which are very resistant to biological degradation. They produce persistent foams which cause unsightly river conditions and interfere with the transfer of oxygen from the atmosphere through the air–water interface. The concentrations of these non-biodegradable surface-active materials in the tidal Thames in the post-war years are shown in figures 50 and 51.

Table 11. Consumption of surface-active materials (synthetic detergents) in the UK, 1949–59 (in thousands of tonnes).

Year	Domestic	Industrial	Total
1949	10.5	2.5	13.0
1950	12.5	2.7	15.2
1951	14.0	3.8	17.8
1952	21.0	5.2	26.2
1953	29.0	6.0	35.0
1954	33.0	6.0	39.0
1955	34.0	6.5	40.5
1956	34.5	6.5	41.0
1957	34.1	6.9	41.0
1958	35.0	7.3	42.3
1959	37.5	7.4	44.9

The rate at which a river absorbs oxygen from the air is measured by the exchange coefficient (see Appendix I), which for the Thames estuary is 5.1 cm/h. Concentrations of detergents in the post-war period caused this to fall to 4.3 cm/h, thus reducing the rate of oxygen entry at the water surface by approximately 16%, and likewise the river's capacity for self-purification by the same amount. Similarly, it was estimated that non-biodegradable detergents in sewage had reduced the efficiency of activated sludge plants by 30%, so that the polluting loads discharged were also correspondingly greater.

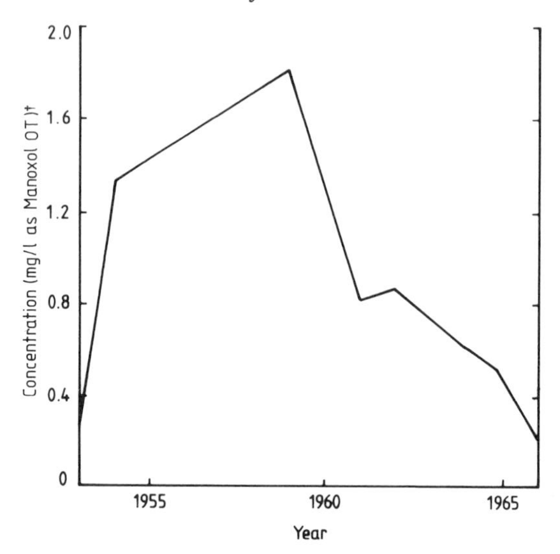

Figure 50. Mean concentration of synthetic detergents in the Thames off the metropolitan outfalls.

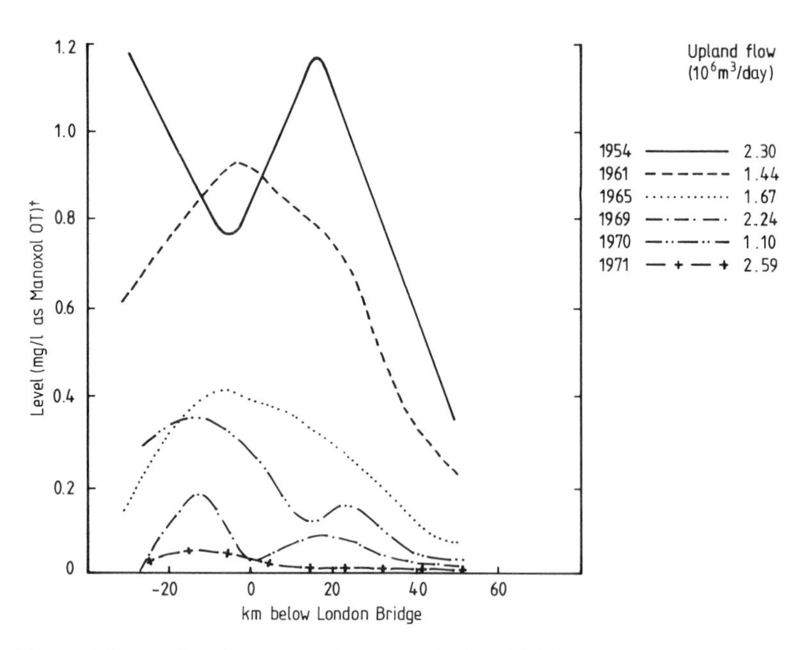

Figure 51. Levels of synthetic detergents in the tidal Thames, 1954–71 (third quarter averages). † Manoxol OT is a surface-active substance. The total surface activity of all detergents in a sample of water can be related to a definite concentration of Manoxol OT.

The difficulties which arose from the use of non-biodegradable ('hard') detergents after the war led to the setting up of the Committee on Synthetic Detergents. This in turn was succeeded by the Standing Technical Committee on Synthetic Detergents in 1957, which included scientists from central government, river authorities and the detergent industry. As a result of its deliberations, the detergent manufacturers, by voluntary agreement, replaced the 'hard' detergents with biodegradable 'soft' products which could be broken down during sewage treatment, and over periods of long retention in the river, i.e. when upland flows were low. It was agreed that the hard detergents would be phased out by 1965 and replaced by the soft ones; figures 50 and 51 show how effective this control was. The composition of synthetic detergents is now controlled by EEC directives (EEC 1973) which lay down requirements for the biodegradability of all detergents used within the member states.

When the condition of the Thames in the post-war years is studied, it becomes apparent that the factor which was ultimately responsible for the deterioration in water quality to completely unacceptable levels was the introduction and widespread use of these hard synthetic detergents. The course taken to remedy the situation is one which exemplifies the uniquely good relations which have always existed between British industrialists and the water industry.

Power Station Effluents

Effluents from power stations (table 12) can affect a river in two ways:

Stack gas washing. In order to minimise atmospheric pollution the stack gases from power stations are scrubbed to remove sulphur dioxide, the washings of which present an oxygen demand when discharged to the river. Battersea power station (6.2 km above London Bridge) exerted a demand of 3 tonnes/day in 1952, and of 6.2 tonnes/day in 1962. At Bankside power station (1.2 km above London Bridge) the scrubbing liquors containing sulphite were aerated in the presence of a catalyst (a manganese compound), and the final load on the river was 1.2 tonnes/day in 1952, and about 2 tonnes/day in 1962.

Thermal pollution. In addition to causing chemical pollution, power stations use immense volumes of cooling water and thereby introduce thermal pollution in the water which is returned to the river at a higher temperature. From mathematical modelling studies it has been found that there has been a rise in the temperature of the river of about 3°C (see figure 52), although it could be much higher in very localised areas due to thermal plumes.

Increased temperatures not only make a river unsuitable for sensitive organisms such as migrating fish, but also decrease the concentration of

Table 12. Electricity generating stations on the Thames Tideway.

Power station	Position below London Bridge (km)	Discharge point (north or south bank)	Heat rejected (10^6 MJ)			
			First quarter	Second quarter	Third quarter	Fourth quarter
Barnes	−18.7	S	0.53	0.15	0.032	0.19
Hammersmith	−15.2	N	1.14	0.70	0.31	0.46
Fulham	−9.8	N	40.7	29.5	30.3	43.3
Lombard Road	−9.5	S	1.65	1.12	0.60	0.82
Battersea	−6.2	S	38.5	30.8	27.1	36.5
Bankside	−1.2	S	7.60	5.02	4.45	4.95
Stepney	4.3	N	7.60	5.79	2.44	5.15
Deptford West	7.2	S	23.0	18.8	17.0	23.3
Deptford East	7.2	S	10.6	9.5	9.28	9.07
Blackwall	12.3	N	0	0	0.84	2.94
Woolwich	16.0	S	8.85	7.76	5.63	6.38
Barking	20.3	N	57.1	46.2	42.7	57.0
Littlebrook	33.2	S	29.1	24.6	15.7	22.8
Gravesend	45.0	S	0.98	0.75	0.49	0.46
Total			227	181	157	213

dissolved oxygen that the water can maintain. The saturation value, C_s, is related to temperature and salinity by the expression (Gameson and Robertson 1955);

$$C_s = \frac{475 - 2.65\,S}{33.5 + T},$$

where S is the salinity (g/1000 g), and T is the temperature (°C). For the 3 °C temperature rise above 20 °C caused by heated effluents from power station, C_s will be diminished by 5.4%, which reduces not only the quantity of oxygen that can be held in solution, but also the value of $(C_s - C)$, and hence the rate of solution of oxygen (see Appendix I).

Thus the growth of riverside power stations using direct water cooling was another factor in reducing the self-purification capacity of the Thames.

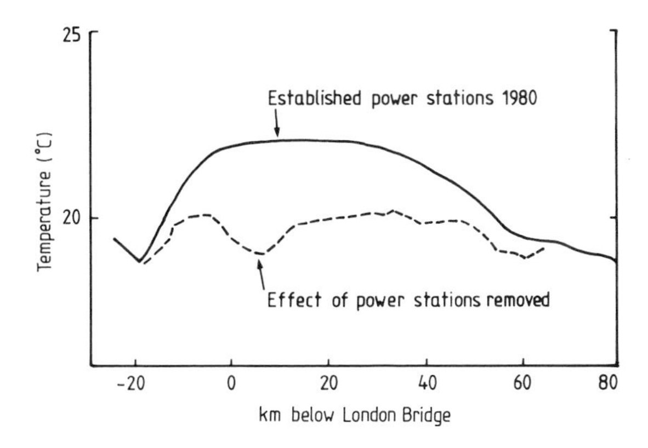

Figure 52. Temperature profile of the Thames estuary, showing the effect of established directly cooled power stations under third quarter minimum flow conditions, 1980 (calculated).

Regionalisation of Sewage Works

On 9 December 1920 representations were made to the then Prime Minister, David Lloyd George, by the London County Council concerning the difficulties in administering the expanding local government services, when the latter were under the control of a multiplicity of local authorities. As a result, a Royal Commission on the Local Government of Greater London was appointed on 24 October 1921, with the following terms of reference:

> . . . to enquire and report what, if any, alterations are needed in the local government of the administrative County of London and the surrounding

> districts with a view to securing greater efficiency and economy in the administration of Local Government services and to reducing any inequalities which may exist in the distribution of local burdens as between the different parts of the whole area.

Amongst other matters there was unanimous agreement that some modification was required in the administration of drainage services, and as a result it was agreed that the Ministry of Health should arrange for an enquiry into the technical aspects of the matter (Taylor *et al* 1935).

The multiplicity of treatment works developed by individual local authorities was noted, together with a projected population growth to 11.3 million in 1971, and possibly 15 million later. The report considered alternative procedures for future policy, and two schemes were considered:

(1) to discharge sewage to sea;

(2) to attempt regionalisation, by treating the sewage from the 182 works at (*a*) 23 or (*b*) 10 regional sewage disposal works.

The former scheme, which had been considered 70 years or so previously was revived, although with different disposal sites in mind. Consideration was given to the following:

Disposal area	Distance from Charing Cross
1(*a*) Havengore Head	69 km
1(*b*) Goring (Sussex)	84 km
1(*c*) Dungeness	101 km

Disposal point 1(*a*) was unacceptable, because of lack of depth of receiving water and possible danger to the shellfish industry. Point 1(*b*) was rejected, because of the use of the south coast as sites for holiday resorts and 1(*c*), whilst hydrologically acceptable, was thought to be too expensive at an estimated cost of £110 million.

It was therefore considered that scheme 2(*a*) would be selected, but only as a stepping stone to the preferred scheme 2(*b*), at an estimated cost of £50 million. The report concluded that:

> . . . consideration should be given to a scheme whereby the whole of the area would be served by ten, or fewer, centralised disposal works.

The ten regions were to comprise the eight areas listed in table 7, together with Beckton and Crossness.

The Mogden Scheme

In 1928 the Middlesex County Council was concerned about future plans for sewage disposal in the west of the county. After an investigation (Watson

1929), Parliamentary powers were obtained in 1931 for the construction of a central purification works at Mogden, Isleworth. This works would discharge directly to the Tideway and replace 27 smaller works, many in a state of obsolescence; some discharged to the Tideway, some to the Upper Thames, and others to the tributaries (table 13). The original Mogden works was opened in 1935 (figure 53). The sewerage system was designed to deal hydraulically with a flow of nearly 3×10^6 m^3/day, and involved the laying of about 120 km of main intercepting sewers, varying in diameter from 0.53–3.9 m.

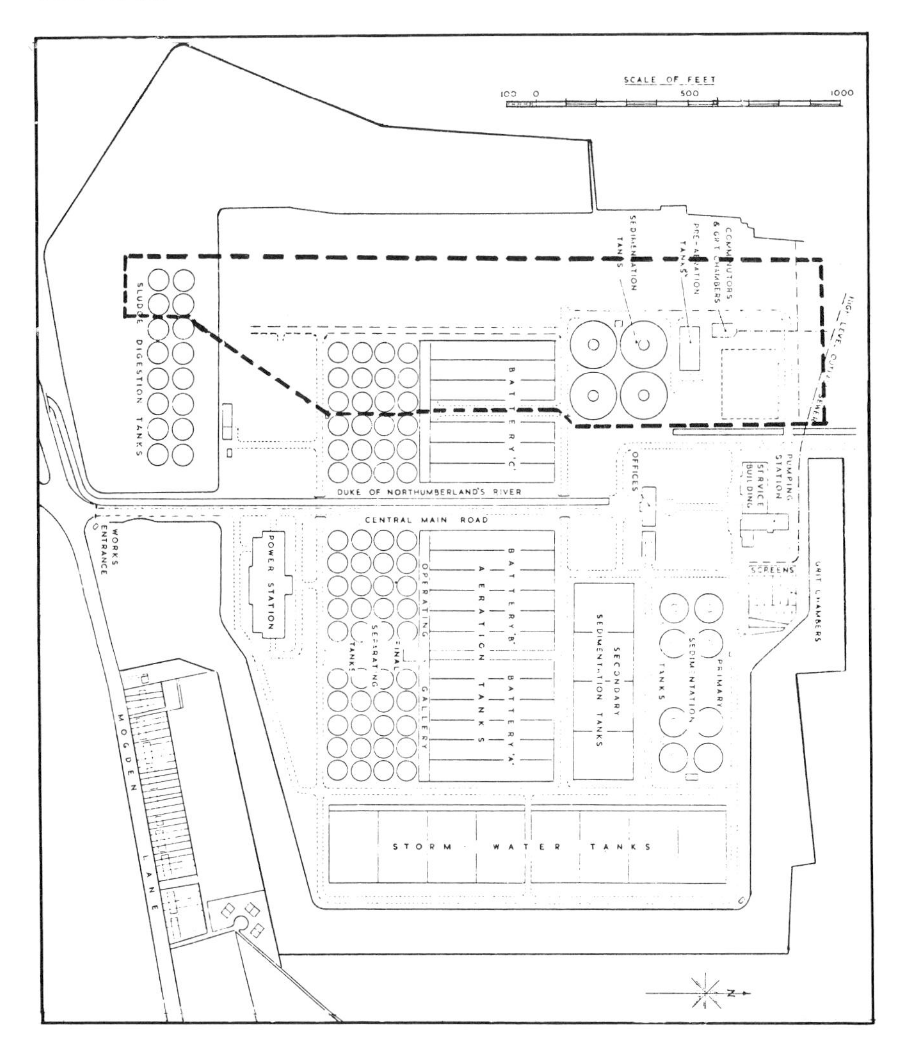

Figure 53. Mogden purification works. The 1962 extensions are enclosed within the broken line.

Table 13. Sewage disposal works of West Middlesex in 1928 (UD, urban district; RD, rural district; MB, metropolitan borough; S, separate; PS, partially separate).

Local authority	Number of works	Sewerage system	Dry weather flow (m^3/day)	Population	Receiving stream
Brentford and Chiswick UD	2	PS	11 500	58 300	Tidal Thames
Ealing MB	4	PS	15 000	100 000	Tidal Thames and Brent
Feltham UD	1	S	910	7500	Crane
Hampton UD	1	S	1770	13 000	Non-tidal Thames
Hayes UD	1	S	2270	13 500	Crane
Harrow UD	1	PS	2150	14 000	Yeading Brook
Hendon UD	1	PS	12 700	89 000	Brent
Hendon RD	3	S	4000	28 600	Yeading, Edgware and Kenton Brooks

Ruislip and Northwood UD	1	PS	2270	10 000	Pinn (non-tidal Thames)
Southall and Norwood UD	1	S	8700	35 000	Brent
Staines UD	1	S	1590	8000	Ash (non-tidal Thames)
Sunbury UD	1	S	3020	6500	Soakaway to subsoil
Teddington UD	1	S	5910	24 000	Tidal Thames
Twickenham MB	1	S	7000	38 000	Tidal Thames
Uxbridge UD	1	PS	3600	14 500	Soakaway to subsoil
Uxbridge RD	2	PS	4090	19 500	Pinn and Yeading Brooks
Wealdstone UD	1	PS	5520	36 000	Wealdstone Brook
Wembley UD	1	PS	5660	28 300	Brent
Total	27		107 900	609 700	

At the time of construction the Mogden works must have been unique, comprising preliminary, primary and secondary treatment, the last by the then comparatively novel activated sludge process using diffused air and providing nine hours' retention in the plant to achieve a substantial degree of oxidation of ammonia. All primary sludge was digested, eventually to provide fuel to run the whole works. The quality of effluents from the Mogden works and the average of the previous obsolescent works it replaced, are given in table 14.

Table 14. Quality of effluent from Mogden and the previous obsolescent works (Sheldon 1979).

	Effluent quality (mg/l)			
	Suspended solids	BOD	Ammonia nitrogen	Organic nitrogen
Previous works' effluent	31	22	19	3
Mogden effluent	6.0	5.7	10	2

However, the achievement in improved effluent quality shown in table 14 would not have produced an improvement in the condition of the Tideway commensurate with the cost of the works. In fact the original survey had been one of planning for future conditions. Between the time that the data of table 13 were collected and the completion of the works in 1935, the dry weather flow had, as predicted, increased from 108 000 to 249 000 m^3/day (Townend 1962). Even assuming that the obsolescent works could have maintained their effluent quality with the increased flow, the pollution loads introduced into the river system would have been much higher, as shown in table 15.

Table 15. Loads discharged from Mogden and from smaller works, 1938.

	Flow (m^3/day)	Load (tonnes/day)		
		BOD	Ammonia nitrogen	Organic nitrogen
Mogden	249 000	1.42	2.49	0.50
Obsolescent works	249 000	5.48	4.73	1.00

By the use of mathematical models (see Appendix IV) it is possible to determine the effect of the regionalisation programme at Mogden on the levels of dissolved oxygen in the Tideway. Figure 54 shows the effect under worst conditions of third quarter minimum flow, giving dissolved oxygen levels:

(*a*) if the existing works had been retained and Mogden had not been built;
(*b*) with Mogden in operation; and
(*c*) if all effluents had been taken to sea (p 74).

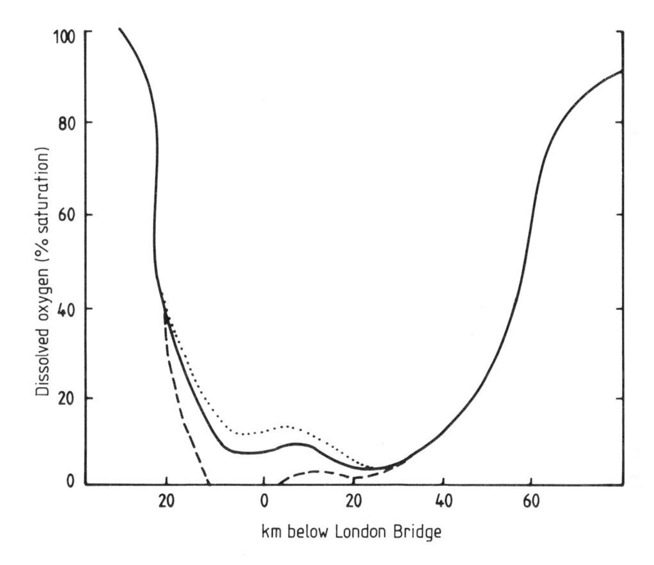

Figure 54. Dissolved oxygen concentration, minimum flow conditions (third quarter 1935). Broken curve, if existing works had been retained and Mogden not built; full curve, Mogden in operation; dotted curve, all effluent to sea.

It can be seen that even if the multiplicity of smaller works had been retained and operated efficiently, there could have been conditions when the river would have become anaerobic for about 14 km. This prediction, however, probably underrates the polluting effects which would have arisen had Mogden not been built. The model assumes that the 27 works continued to produce the quality of effluent shown, even with the increased flow. It is most unlikely that this would have been possible, since the filters would probably have been subject to failures such as 'ponding', apart from mechanical breakdowns and staffing problems. In fact, the single works at Mogden had many advantages, such as greater economy of labour and lower operating costs, and any plant failures could be speedily rectified by staff on site. In addition, the system discharged all effluents to the Tideway, which was able to accept them and remain of satisfactory quality. Previously, many of the discharges had been made to tributaries with insufficient flow to be capable of receiving them. These waterways improved considerably when Mogden was established.

As a result of suburban development after the war, the sewage flow to Mogden inevitably increased (figure 55), and, as was to be expected, the plant became unable to cope with the higher loads imposed upon it, and the effluent quality deteriorated, especially as a result of high loading from storm flows (figure 56). The condition was at its worst in 1962, but by that time the additional treatment plant had been built, which included a primary sedimentation and an activated sludge plant (figure 57). The improvements in quality since 1962 are illustrated in figure 56. The effect of the effluent on river quality was obtained by mathematical modelling (figure 58), and it was found that under third quarter minimum flow conditions the river would have been anaerobic for long distances, but fortunately the storm flows to the works seldom coincided with minimum upland freshwater flows. The average conditions in the third quarter of 1962 are shown in figure 58.

The quality of the upper Tideway was affected by the discharges from Acton sewage works, from which virtually untreated sewage was discharged, and also to some extent by the metropolitan sewage works' effluents. Therefore the improvements effected by the installation and maintenance of Mogden works was obscured by the effects of these other effluents, but there is no doubt that had Mogden not been maintained to supply good-quality effluent,

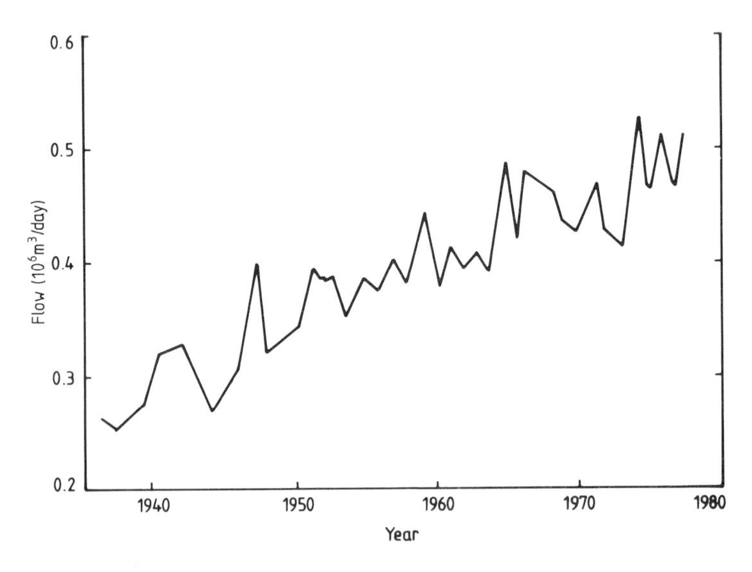

Figure 55. Flow of sewage to Mogden works.

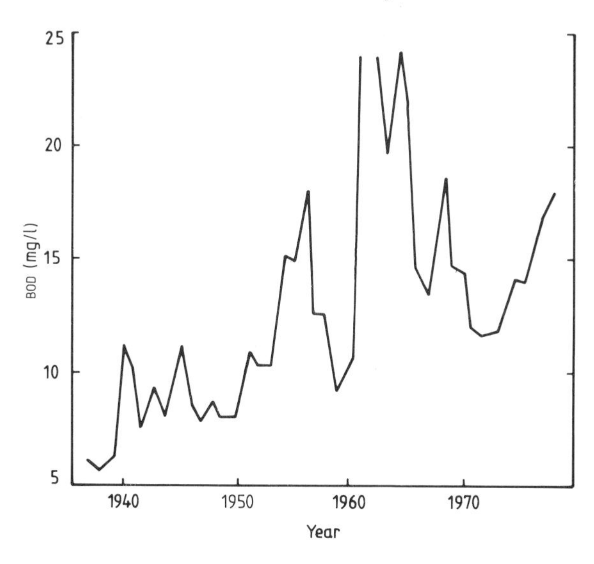

Figure 56. BOD of average effluent from Mogden works.

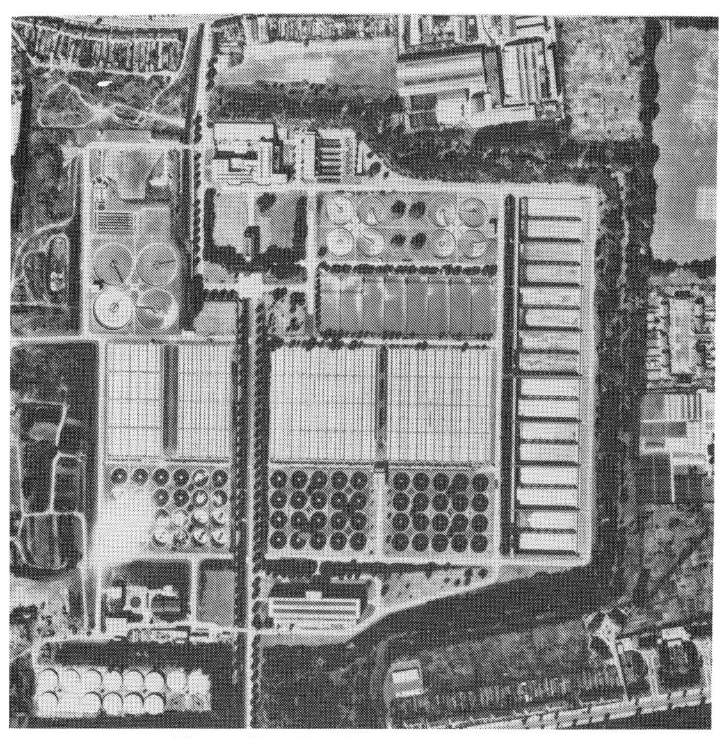

Figure 57. The Mogden works in 1980.

there would have been a deterioration in the upper Tideway as bad as that produced by Beckton and Crossness. The effects of other effluents can be disregarded as having minimal effects compared with those described.

A further reorganisation scheme, that of East Middlesex, was adopted in 1963 when the new Deephams works, discharging into the River Lee, replaced 11 smaller works. Otherwise, the regionalisation plan for London has been only partially implemented.

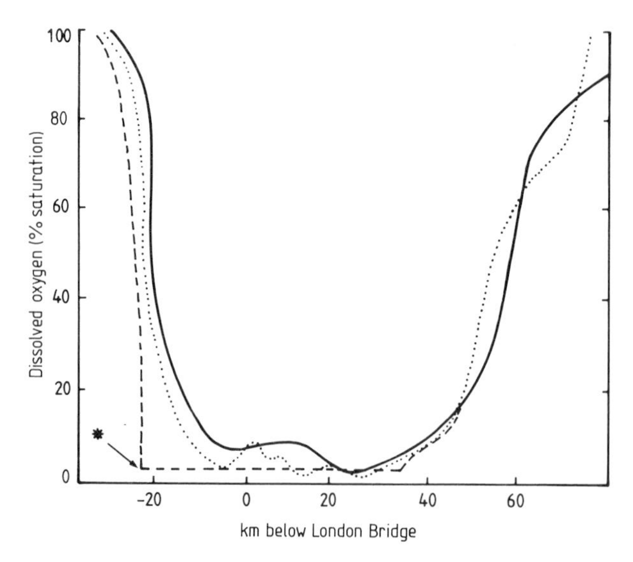

Figure 58. Dissolved oxygen concentration (third quarter averages). *Model unreliable below 4% saturation. —— 1935, ····· 1962, ----- 1962 minimum flow conditions (0.77×10^6 m^3/day) modelled.

Condition of the River in the Early 1950s

The deterioration of the Tideway reached its worst in the early 1950s when long stretches of the river were devoid of oxygen (figure 59). In these anaerobic regions no aquatic life was possible. In such a river system, when the reserves of dissolved oxygen become seriously depleted and fall below about 5% saturation, the micro-organisms that normally decompose any polluting organic matter can change, and use is made of oxygen combined in salts such as nitrate, the latter being decomposed to produce elementary nitrogen, an innocuous gas which passes into the atmosphere without giving offence (see Appendix I).

When nitrate reserves are exhausted, however, anaerobic conditions prevail and oxygen is removed from sulphate, generating sulphides, which are not only toxic to all forms of life, but can also give rise to hydrogen sulphide,

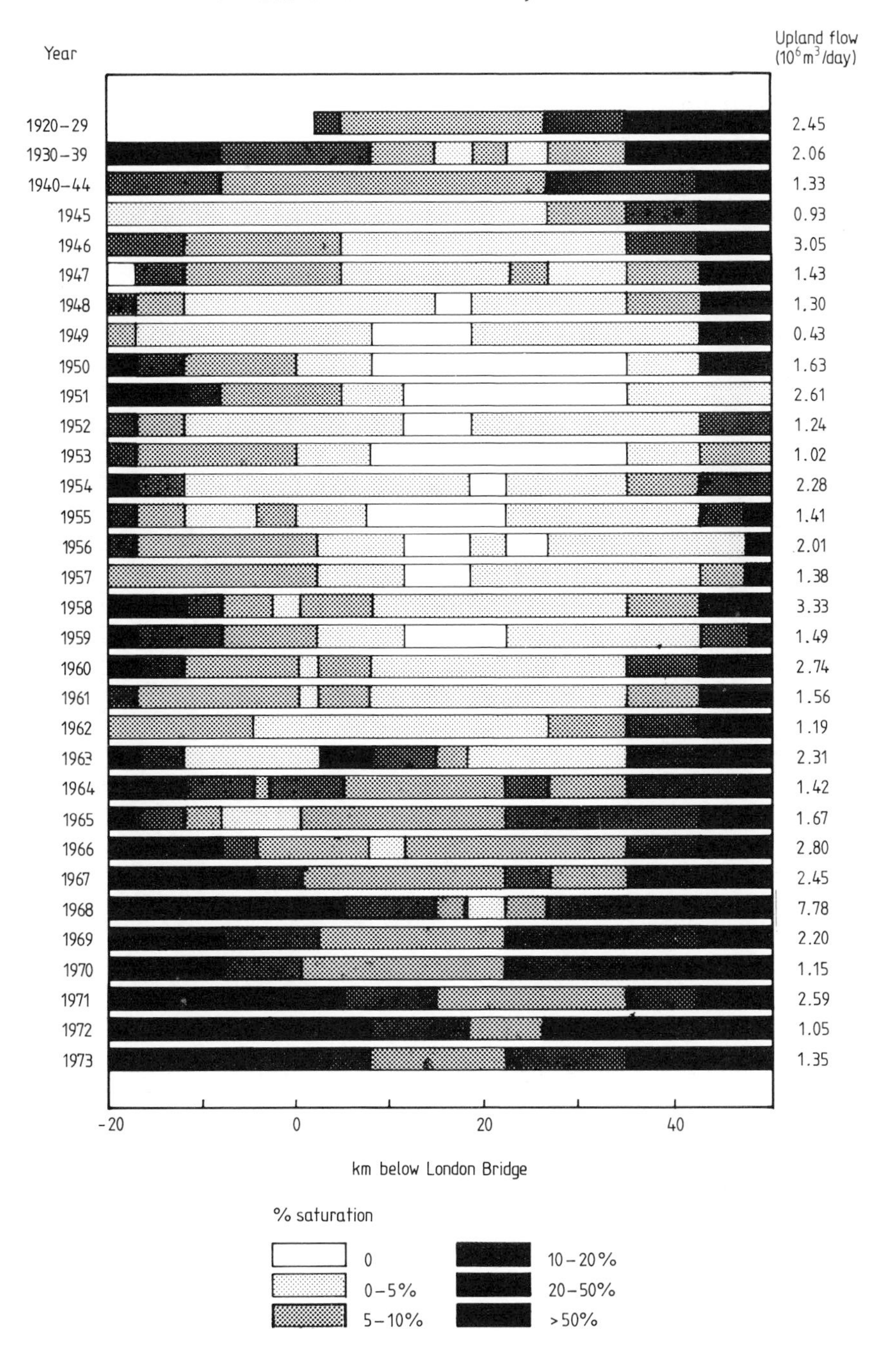

Figure 59. Concentration of dissolved oxygen (% saturation) at high water (third quarter averages), 1920–73.

a very toxic and corrosive gas. Nitrates therefore act as a safeguard against sulphide formation, and only in exceptional circumstances can they coexist, as can be seen from figure 60.

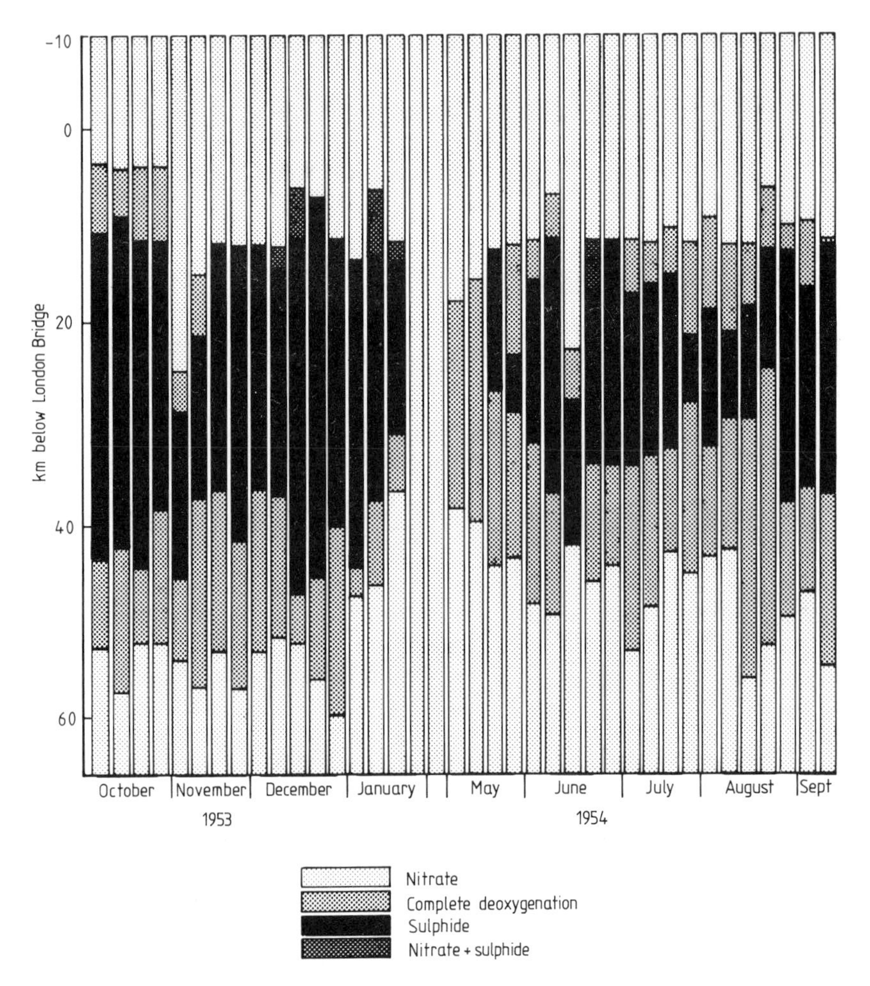

Figure 60. Incidence of nitrate, complete deoxygenation and sulphide at half tide, 1953–54.

In the early 1950s sulphide was present in the Thames, as shown in figures 61 and 62. The dissolved hydrogen sulphide can be displaced from solution by disturbance of the surface by wave motion, for example, or by entrainment in bubbles when other gases are generated, such as nitrogen or carbon dioxide. Hydrogen sulphide has an offensive odour and tarnishes metals and their

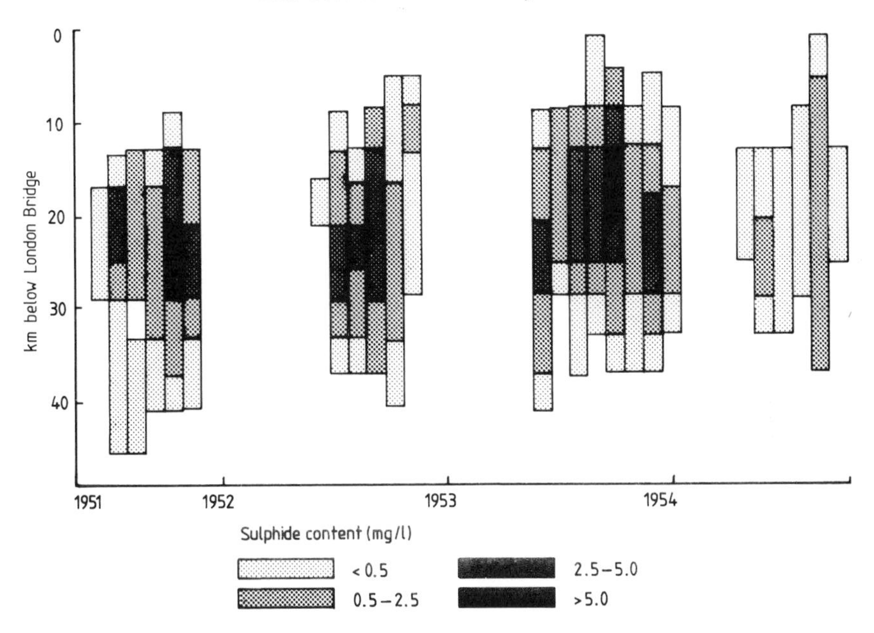

Figure 61. Lengths of river containing sulphide.

compounds. In the early 1950s at sites adjacent to the river, concentrations of sulphide of up to 2 ppm were found in the atmosphere, and copper, brass and iron blackened almost immediately. Much of the paint used on ships' hulls were lead-based, and this was also rapidly blackened, as was the lead glaze of baths in riverside areas.

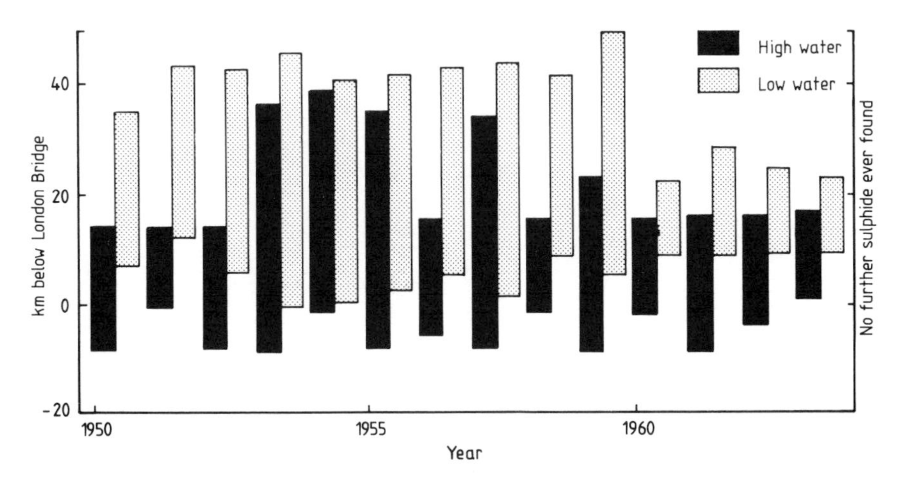

Figure 62. Lengths of the Thames containing sulphide on any one occasion.

The very poor state of the Thames coincided with the Festival of Britain in 1951, when many foreign ships were to be moored there. Attempts were made to improve conditions by the use of chlorine at Beckton with the object of sterilising the effluent so that its decomposition would begin farther downstream where more dissolved oxygen was available. This work was carried out during July–October 1951, but it produced little or no improvement. It became evident that the polluted condition of the river which flowed through the capital could not be allowed to continue.

The Second Restoration, 1950–80

The measures that were taken to restore the polluted Tideway resulted from a unique combination of enthusiasm of responsible bodies and devoted scientific investigation.

The Port of London Authority was responsible, *inter alia*, for the condition of the tidal river and creeks, but its powers were limited. In 1948 the River Boards Act set up river boards to exercise control of river management in their particular catchments. The boards were later provided with legislative powers to issue consents for discharges to streams, by the Rivers (Prevention of Pollution) Acts of 1951 and 1961. These powers were not available to the Port of London Authority, and in respect of the discharges from Beckton and Crossness the Authority had few, if any, powers. The Act for the Better Local Management of the Metropolis of 1856 was still extant, and required only that the sewage effluents from these works did not return to the metropolis. Despite their difficulties, the Port Authority's officers worked assiduously—mainly by persuasion—towards ensuring improved effluent qualities, and eventually some industrialists, seeing improvements being made, enthusiastically contributed their support of their own volition.

The Greater London Council and its predecessor at this time, the London County Council, could have carried out merely what was required by statute. Even at the time when the slogan 'There are no votes in sewage!' was frequently heard, this was not their attitude, no matter which political party was in power; the tradition of providing the best lifestyle for Londoners seems to have been the main driving force of the council. The aim was first to provide London with a waterway free from nuisance, and then to have a river affording amenities to the residents. Many engineers including Messrs W P Warlow and E H Vick, together with the Chemist in Chief, Mr C J Regan (figure 63), and their senior chemist of the outfall works, Mr G S Clements, were largely involved.

A river cannot be cleaned, even with the best of goodwill, solely by local government and regulatory bodies, and the Thames rehabilitation undoubtedly owes a great deal to the study, probably the most comprehensive of any made on a river, by the scientists of the Water Pollution Research Laboratory (now the Water Research Centre) at Stevenage.

Figure 63. Colston James Regan, Chemist-in-Chief, London County Council, 1941–54.

The Thames Survey Committee

In 1947 the Chairman of the Port of London Authority, Lord Waverley, requested of the Secretary of the Department of Scientific and Industrial Research that the Water Pollution Research Laboratory should assist in an investigation of the causes of silting in the Thames estuary, especially in the neighbourhood of Barking, to ascertain in particular whether the deposition of mud was being affected by the discharge of polluting matter. As a result the Thames Survey Committee was established, its members including scientists and engineers from the Port of London Authority, London County Council and central government, as well as consultants in this field of public health engineering. The Laboratory's investigation started in May 1949. Because the siltation work was eventually undertaken by the Hydraulics Research Station, using large, purpose-built physical tidal models, the Thames Survey Committee turned its attention mainly to the polluted state of the river, especially as the evolution of sulphide in the dry summer months was giving the public reason to complain frequently. The Laboratory therefore extended its work, and a study lasting over ten years was undertaken into the causes of, and remedies for, the pollution of the Tideway. Their report, *The Effects of Polluting Discharges on the Thames Estuary* (Water Pollution Research Laboratory 1964), was the outcome.

The work began with special surveys to supplement those initiated by Dibdin in 1893, and which had been continued since then by the London County Council, and were considered to be more complete and detailed than for any other estuary in the world. This was in itself remarkable, since the London County Council and its predecessors did not have the control of, nor the responsibility for, monitoring the river—it was indicative of the attitude which they adopted towards the restoration of the Thames in the interests of their ratepayers.

The Laboratory studied in detail the rates of degradation of polluting matter, the concentrations of dissolved oxygen under which various types of biodegradation reactions proceed, and the mechanism of introducing and replenishing dissolved oxygen. The effect of factors such as temperature, the upland flow of fresh water, sea-water penetration and mixing characteristics were also studied. It was understood that the study would be very labour-intensive, but the Committee recognised that the work would benefit not only the Thames, but would also be of value in the rehabilitation of other estuaries. The published work of the Laboratory already referred to, enabled a mathematical model to be constructed which is used with little change today, and also enabled an assessment to be made of the effect of the position of the discharges on the quality which would be required of them, no matter whether their effect be of the nature of oxygen-consuming, non-biodegradable or thermal pollution. The Laboratory found that the reactions involving biodegradation of polluting substances in the estuary could be represented by the kinetics of the kind shown in Appendix I.

This study recognised the importance of ammonia and nitrate on the quality of water in an estuary. It was shown that ammonia discharged in effluents will make a heavy demand upon the dissolved oxygen resources during its oxidation in the Tideway. On the other hand, the nitrate produced by this oxidation can, as has previously been stated, provide a reserve of oxygen to prevent the onset of putrid conditions when dissolved oxygen levels are low. The Thames Survey Committee considered that the discharge of nitrate can never affect an estuary adversely†, and from this came the first

† It must be pointed out, however, that the desirability of the presence of nitrate applies only to estuaries or waters not used for potable supply. In rivers used for the latter purpose, the intake levels must be kept preferably below the World Health Organisation desirable limit of 50 mg/l NO_3'' and certainly below the mandatory limit of 100 mg/l NO_3''. Nitrate in water can give rise to the condition in bottle-fed infants known as methaemoglobinaemia ('blue' babies), and is suspected of having a carcinogenic effect due to reduction to nitrite and reaction to form N-nitrosamines in the human gut.

concept of an environmental quality objective (EQO). It was known that the estuary would be kept free from nuisance provided that nitrate was present, and the Committee concluded that this might form the basis of a suitable standard, provided that a margin of safety was allowed. Reserves of nitrate begin to be used up when dissolved oxygen falls to about 5% saturation. The Committee therefore suggested a standard for the Tideway of a minimum of 10% saturation with dissolved oxygen in all places and at all times, thus providing a margin of safety against nuisance (hydrogen sulphide formation) of 5%.

The concept of an environmental quality objective, recognising that it is more important that the environmental quality as a whole should be maintained, rather than that 'fixed emission' standards should be imposed, is typical of the attitude taken in the United Kingdom and has generally led to successful results.

The Pippard Committee

On 4 April 1951 the then Minister of Local Government and Planning set up the Departmental Committee on the Effects of Heated and other Effluents and Discharges on the Condition of the Tidal Reaches of the River Thames. The Committee was to sit under the chairmanship of Professor Pippard, with the following terms of reference:

> . . . to consider the effects of heated and other effluents and discharges on the condition of the tidal reaches of the River Thames, both as at present and as regards any proposed new developments in the area, and to report.

At the time of the setting up of the Pippard Committee, it was realised that not enough was known about the factors responsible for the condition of the estuary to formulate a policy for its management, and that a prolonged study would be necessary. Advantage was taken, therefore, of the work being carried out under the auspices of the Thames Survey Committee by the Water Pollution Research Laboratory, and this was a further reason for the latter to broaden their work from the causes of siltation to a study of the sanitary condition of the river. The two committees worked together very closely, nearly all members serving on both, and the scope of the research programme was extended to cover the terms of reference of both. Generally the Thames Survey Committee remained responsible for the research programme, whilst the Pippard Committee collaborated by obtaining information on the various discharges to the river.

Three years before the publication of the Water Pollution Research Laboratory Report, the Pippard Committee had gathered sufficient information to formulate its own report (Ministry of Housing and Local Government 1961). This contained conclusions of two types; some were of a general nature, dealing with the overall condition of the estuary and the aims and methods of future management, while others dealt with specific issues such as the effects of particular effluents on estuarine quality. The Committee came to the following general conclusions:

(1) In the early 1950s the Thames estuary was in a badly polluted and often offensive state as far downriver as Gravesend, and there were frequent examples of complete deoxygenation with the evolution of hydrogen sulphide.

(2) The polluting loads of discharges were based upon both their carbon and nitrogen contents, and Pippard subscribed to the views of the Thames Survey Committee that whereas discharges of oxidisable carbon are always polluting, the effects of those containing oxidisable nitrogen are more complex. Using the results obtained by the Water Pollution Research Laboratory, the Pippard Committee was able to assess the effect of significant discharges on the quality of the water in the river.

(3) The Committee considered that it was of the utmost importance that the quality of the estuary should be satisfactory in all places and at all times, but that uniform quality standards need not necessarily apply. Therefore, in assessing the standard of consent for the discharge of an effluent, account should be taken not only of quality but of quantity, and each discharge should be treated on its own merits, based upon its effects on the quality of the estuary.

(4) The Committee commented on the fact that the authority discharging the greatest pollution load, the London County Council, was also responsible for monitoring the Tideway. It saw no reason to change this, but considered that in future, additional assessments should be made by the Water Pollution Research Laboratory when remedial measures had been put in hand.

(5) The Committee considered that the Port of London Authority should be given the statutory pollution control powers possessed by river authorities, to impose conditions for the discharge of effluents, and likewise to dischargers, the right to appeal against these conditions if they were dissatisfied.

(6) The Committee's recommended aim was to prevent offensive conditions and to provide a margin of safety. It concluded that this would be achieved provided that conditions allowing the generation of hydrogen sulphide were obviated by ensuring that there was always a detectable quantity of dissolved oxygen or nitrate present in the water.

The Committee believed that its aims would be met, provided that the polluting loads then being discharged (1961) were reduced by 75%. In its opinion, to raise the quality of the Tideway in respect of the dissolved oxygen concentration to a level which would permit the establishment of migratory fisheries (salmon and sea trout), would require very high standards of effluent, the cost of which would be out of all proportion to the benefits gained. The Water Pollution Research Laboratory had concluded that to re-establish the salmon population would require that in nine years out of ten, the river should have a concentration of dissolved oxygen of not less than 30% saturation in April and May, the months when the salmon smolts migrate to sea.

The Pippard Committee therefore made the following observations and suggested these specific recommendations for improvement:

(*a*) The sewage effluents which had the greatest effect on the Tideway were those from Beckton, Crossness, Mogden, West Kent and Acton. The first four contributed 90% of, and Beckton 50% of, the total sewage effluent discharged to the river. Mogden and Acton, because of their positions, should therefore be required to produce high-quality effluents. It was essential that the effluents from Beckton and Crossness should be improved.

(*b*) Storm sewage overflows contributed about 1.5% of the total pollution load, but their effect was greater than this implied, because they were discharged further upstream than the largest sewage works (and often as a large load over a short period).

(*c*) Industrial loads amounted to 9% of the total in 1952–53, but by 1961 this had been reduced to 3%.

(*d*) The introduction of synthetic detergents in and after 1949 was followed by a marked increase in the length of the zone of river showing a complete absence of dissolved oxygen.

(*e*) Heated effluents from power stations had caused temperature increases of up to 5 °C in the estuary, and this had accelerated the consumption of oxygen. Additional discharges of heated effluents should be avoided in the most polluted reaches, and any effluents entering these reaches should be aerated.

(*f*) The Thames showed no improvement when London's population decreased in the years 1939–45, and this could not be accounted for by events that were known to have occurred.

(*g*) New polluting discharges above Lower Hope Reach (50 km below London Bridge) should be avoided, and proposals for discharges even below this point should be carefully considered.

Probably the two most important developments in the improvement of the condition of the Tideway which followed the Pippard Report were the rebuilding of Beckton and Crossness (p 103 *et seq.*) and the granting to the Port of London Authority effective pollution control powers. The Port of London Act 1968 granted to the Authority those powers already exercised by river authorities under the Rivers (Prevention of Pollution) Acts of 1951 and 1961.

The Royal Commission on Environmental Pollution (Third Report)

In the decade following the publication of the Pippard Report, public opinion was changing towards the need for better environmental standards, and the

Third Report of the Royal Commission on Environmental Pollution, under the chairmanship of Sir Eric (now Lord) Ashby, in 1972 provided what can be regarded as a blueprint for estuarial management. Thus although remedial work was in hand on the recommendations of the Pippard Committee, the report of the Royal Commission enabled the situation to be reviewed and rationalised.

The Commission expressed concern about the polluted state of some estuaries, and the absence of appropriate legislation to deal with it, such as was available in the case of non-tidal rivers. It found the public to be confronted by two attitudes. (i) An emotional approach which, recognising that polluting discharges can damage or destroy shellfish, birds and fish, considered that all such contamination should be stopped to reverse the process of destruction. (ii) The opposing view that tended to underrate the damage done and stressed that discharging sewage and industrial effluents to estuaries considerably reduced the costs which would otherwise throw a very heavy burden on industry and the local community, and could involve the risk of unemployment.

The Commission adopted an intermediate approach, and considered that there was a practical limit to the burden which should be placed upon the community for the abatement of pollution of estuaries, this limit being defined as the point beyond which the marginal cost of abatement exceeded the marginal cost of the damage done. Since such costs were not readily available, a more pragmatic approach was desirable, and the Commission recommended two methods of controlling the discharges to the estuary.

(1) Pollutants not rendered harmless by natural processes, and which accumulate in benthic mud or in living organisms, should be removed from effluents before they are discharged.

(2) In respect of biodegradable wastes, by using monitoring data, the quality of the estuary should be maintained by the establishment of 'pollution budgets'. These would measure the maximum polluting load the estuary could accept, if good quality was to be maintained. The general aims of management should be:

(i) to exploit the estuary for waste disposal up to a level which does not endanger aquatic life, or transgress the standards of amenity which the public need and are prepared to pay for;

(ii) to ensure, by controlling quality standards, that exploitation of the estuary does not exceed this level; and

(iii) the policy of planned pollution budgets would require close cooperation between planning authorities and river authorities (or their successors), and there should be consultation on any relevant plans.

The Commission suggested two simple biological criteria for estuarial management: the estuary should be able (*a*) to support on the mud bottom the fauna essential for sustaining sea fisheries; and (*b*) it should allow the passage of migratory fish at all states of the tide. It also recommended that whilst the government should integrate overall pollution control of estuaries with a national policy for waste disposal, the executive responsibility for controlling the pollution of a particular estuary should rest with a single authority. The Commission proposed amendments to existing legislation to give regional water authorities greater control over all discharges to sewers, rivers, estuaries and coastal waters; these statutory powers now form part of the Control of Pollution Act, 1974. It was recommended that the government should take a lead in helping to reach international agreement for the publication of monitoring data regarding the quantities of certain pollutants reaching the sea. Regional water authorities should be responsible for monitoring discharges; essential substances should be thus examined, and certain typical organisms used as 'indicator' species and monitored.

The Commission recommended that greater effort should be devoted to developing mathematical models (for deriving pollution budgets), and that there should be further research into the toxicity of common pollutants to aquatic organisms; on non-biodegradable substances and their accumulation in and release from benthic sediments; and on the effects of trace amounts of organochlorine and organomercury compounds on photosynthesis by marine phytoplankton. The members of the Commission could not, however, agree on a policy for charging for the control of pollution.

Application of Royal Commission Recommendations

To some extent the work of the Thames Survey and Pippard Committees had anticipated the findings of the Royal Commission, the Commission's Report observing that 'the improvement brought about in this [the Thames] estuary provides an excellent example of how the scientific study of an estuary can be used to pinpoint the action required to clean up pollution'. As a basis for estuarine management, however, the report did lay down clear guidelines for targets for water quality. These again followed the philosophy adopted by the Thames Survey Committee, namely, that it was necessary to ensure that estuarine quality was satisfactory in all places at all times, that is, to set an environmental quality objective (EQO) for the estuary, rather than to adopt fixed emission standards for all discharges. This typical United Kingdom approach has since been in conflict with some EEC directives based upon fixed emission standards.

In order to adopt the Commission's recommendations in the case of the tidal Thames it was therefore necessary to take the following steps:

(*a*) to control and remove, as far as possible, toxic, non-biodegradable substances from effluents;

(*b*) to accept the need for an EQO permitting passage of migratory fish at all states of the tide;

(*c*) to establish a pollution budget to allow the EQO to be achieved; and (*d*) to monitor the estuary to confirm that the EQO was being met, and to examine the biota to ensure that the benthic organisms required to support sea fisheries were present.

(*a*) *Control of toxic, non-biodegradable substances.* These substances can enter the Thames Tideway either in effluents discharged from sewage and industrial works, or in the sewage sludge disposed of at the spoil ground in the Barrow Deep (110 km below London Bridge). In sewage effluents such materials are derived either from normal domestic waste, or from trade discharges to sewers. Little can be done to control the presence of these toxic materials in domestic sewage, householders having the right to have their sewers connected directly to the main drainage system so that obviously no monitoring of their discharges is practicable. Fortunately, the levels of such substances are low since they are derived from such sources as the corrosion of galvanised gutters, the zinc coating of which may contain cadmium. In the case of discharges from trade premises, however, the situation is rather different. Legislation—such as the Public Health Act, 1936, the Public Health (Drainage of Trade Premises) Act, 1937, the Public Health Act, 1961, and the Control of Pollution Act, 1974—requires that traders apply to a water authority for a consent to make a discharge, which can have conditions written in, such as limitations of concentrations and quantities of toxic substances, temperature, pH (see Appendix III). These consents can be reviewed and changed if necessary, after periods normally of not less than two years. In this way it has been possible to limit the concentrations of toxic, non-biodegradable substances in sewage, and therefore in the effluents discharged to the tidal river and in the sludge disposed of to sea. Similar control can be exercised under other legislation (such as the Port of London Acts, 1964 and 1968, and the Control of Pollution Act, 1974) to limit by the consent procedure the concentrations of these substances in industrial effluents discharged directly into the tidal Thames. The present levels of non-biodegradable substances in the Tideway are shown in table 16 and concentrations of metals in fish in table 17. The levels, both in the water and in fish, are considered to be satisfactory.

(*b*) *An environmental quality objective for the Thames Tideway.* The EQO chosen for the Thames Tideway was that which would provide suitable conditions for the passage of migratory fish at all states of the tide. The Water Pollution Research Laboratory in its Report, *The Effect of Polluting Discharges on the Thames Estuary* (1964), suggested that in order to re-establish a salmon fishery in the Tideway it would be necessary to have a dissolved oxygen concentration of not less than 30% in April and May, when the salmon smolts (the most sensitive of this class of fish) migrate to sea.

The Thames Migratory Fish Committee sat from 1973–77, and reported that the requirements for the restoration of a salmon fishery would be a minimum dissolved oxygen concentration of 35% saturation in May, and a

Table 16. Average concentrations of heavy metals, organohalogen and detergent compounds in the Tideway.

Sampling point	Position below London Bridge (km)	Metal concentrations, 1976–79 (μg/l)						Organohalogen compounds, 1975–78 (μg/l)						Detergents, 1975–78	
		Zinc	Copper	Nickel	Lead	Cadmium	Mercury	α-BHC (HCH)	γ-BHC (HCH)	Aldrin	Dieldrin	DDT	PCB as Archlor 1254	Anionic as Manoxol OT (mg/l)	Non-ionic as Lissapol NX (mg/l)
Richmond	−25.8	52	23	20	15	2.3	0.36	2.2	4.5	1.0	2.0	1.3	3.2	0.10	0.01
Isleworth	−24.1	62	19	26	18	2.3	0.25								
Brentford	−22.0	76	25	30	18	2.3	0.67							0.09	0.01
Kew	−20.9	86	31	35	28	2.4	0.29								
Barnes	−17.7	65	23	28	27	2.3	0.32	1.6	3.7	1.6	2.0	2.0	4.8	0.10	0.01
Putney	−11.9	91	29	31	27	2.7	0.22								
Battersea	−7.9	84	45	32	29	2.7	0.44	1.4	2.8	1.3	2.0	2.4	8.6	0.10	0.02
Vauxhall	−4.5	106	43	31	32	3.0	0.85								
Charing Cross	−2.4	117	40	32	38	3.0	0.42	1.1	1.4	<1.0	2.0	2.7	8.2	0.09	0.02
London Bridge	0	96	37	47	39	2.7	0.36								
Cherry Garden	+1.9	84	31	42	31	2.3	0.63							0.08	0.02
West India Dock	+4.7	112	39	43	32	2.7	0.60								
Greenwich	+7.7	118	37	43	31	2.9	0.66	1.1	3.0	<1.0	2.1	3.7	10.2	0.09	0.02
Victoria Dock	+11.4	80	29	41	28	2.8	0.90								
Woolwich	+14.7	85	34	48	34	3.2	0.89							0.11	0.02
Beckton	+18.4	92	28	37	23	2.5	0.39	1.2	4.0	<1.0	1.8	3.6	7.2	0.10	0.02
Crossness	+21.9	79	28	41	20	2.7	0.51	<1.0	2.7	<1.0	1.4	1.6	7.6	0.11	0.01
Erith	+26.6	70	24	37	15	2.0	0.60								
Greenhithe	+34.8	81	21	38	15	2.0	0.59	<1.0	1.7	<1.0	1.4	2.4	5.6	0.08	0.01
Gravesend	+42.5	99	31	34	13	2.0	0.35								
Ovens Buoy	+47.7	67	30	29	11	2.1	0.81							0.07	0.02
Mucking	+53.2	57	19	19	25	2.0	0.30								
Chapman Buoy	+62.5	59	16	22	17	2.0	0.25	<1.0	<1.0	<1.0	<1.0	1.6	3.1	0.06	0.02
Southend	+69.7	40	22	16	18	2.4	0.29								
No. 2 Sea Reach	+77.6	43	17	10	20	2.1	0.39							0.04	0.02
North Oaze Buoy	+80.8	39	14	10	17	2.1	0.46								
NE Mouse Buoy	+93.6	38	16	9	22	2.3	0.93							0.04	0.01
Mid-Barrow LV	+103.7	23	15	8	17	2.1	0.51								
Barrow No. 7 Buoy	+111.0	24	16	10	18	2.1	0.38	<1.0	<1.0	<1.0	<1.0	2.1	3.2	0.04	0.01

Table 17. Concentrations of metals in muscle of sole (*Solea solea*) (mg/kg).

Age	Size (cm)	Zinc				Copper				Nickel				Cadmium				Mercury			
		a	b	c	d	a	b	c	d	a	b	c	d	a	b	c	d	a	b	c	d
0+	4–10	10.6	7.0	17.6	9	1.02	0.24	2.29	8	0.10	ND	0.30	8	0.07	ND	0.19	9	0.03	0.01	0.07	9
1+	11–19	9.1	6.6	20.3	6	0.76	ND	1.72	6	0.22	ND	0.57	6	ND	ND	ND	—	0.02	0.01	0.04	5
2+	20–27	8.0	3.3	20.3	4	0.50	ND	1.43	5	0.17	ND	0.49	4	0.03	ND	0.12	4	0.07	0.06	0.11	4
3+/4+	27	5.5	5.1	6.1	3	0.38	0.23	0.51	3	0.08	ND	0.23	3	0.01	ND	0.03	3	0.02	0.02	0.02	3

a = mean value
b and c = lower and upper levels of range
d = number of samples
ND = none detected

Levels of lead were below limits of detection.
Approximate limits of detection(mg/kg)

Zinc	0.1	Lead	1.0
Copper	0.15	Cadmium	0.05
Nickel	0.20	Mercury	0.002

temperature not exceeding 20°C (Thames Water Authority 1977). The river experiences its worst conditions in the summer quarter (July–September), and statistically the dissolved oxygen level of 35% in May corresponds to a minimum average of 30% for the third quarter of the year. For management purposes, a quarterly minimum average is an inconvenient measurement, and as it corresponds to a 95 percentile value of not less than 10% dissolved oxygen saturation in all parts of the river, this is taken as the standard.

Somewhat ironically, this is precisely the standard suggested by Thames Survey Committee, although it was considered that the level of 30% dissolved oxygen saturation for migratory fish would be unjustifiably expensive to achieve. This standard now relates to the most polluted reaches of the river; in other areas more stringent standards are adopted (p 139).

(*c*) *Establishment of pollution budgets.* Work had been proceeding on upgrading the sewage works at Beckton and Crossness since the early 1950s, before the concept of pollution budgets had even been considered, and at the time of publication of the Royal Commission's Report in 1972 some reconstruction had already been completed. Nevertheless the Report gave an opportunity for the programme to be re-examined. Having established the EQO for the Thames Tideway as a minimum average of 30% saturation with dissolved oxygen, the pollution budget is the amount of dissolved oxygen that can be used in the Tideway before the oxygen curve falls to that level, and hence the pollution load which will require that amount of oxygen for its purification in the river. In figure 92 the area above the 30% dissolved oxygen ordinate is theoretically the pollution budget. In practice, as explained on p 140, the area above the curve is a more appropriate pollution budget.

The mathematical model could be used to determine the effect on the quality of the Tideway of effluents of different standards discharged from existing sewage treatment works, and which of these standards were necessary to meet the pollution budget. There were obviously an infinite number of possible arrangements of suitable effluent qualities, but there were also practical restraints in that some works had already been upgraded to certain standards. The mathematical model showed how the budget could be met, in the circumstances obtaining, with the most cost-effective use of capital.

Crossness had been completed in 1963, but an assessment could still be made of the effluent quality required at Beckton. The model was run using inputs for effluents of quality as set out in the following table.

	BOD (mg/l)	Ammonia nitrogen (mg/l)
(1)	10	10
(2)	10	5
(3)	10	2.5

The model predicted the results shown in figure 64 for third quarter minimum flow conditions, and it can be seen that even with the lowest-quality effluent there was a considerable improvement over the performance of the existing works (third quarter, 1971). However, it was considered that the improvement in river quality which would result from an effluent of quality (3) in the table would justify the cost of a treatment plant of a size capable of producing it, especially as even with this quality of effluent, the EQO required for the Tideway would not quite be achieved. The works at Beckton was designed accordingly.

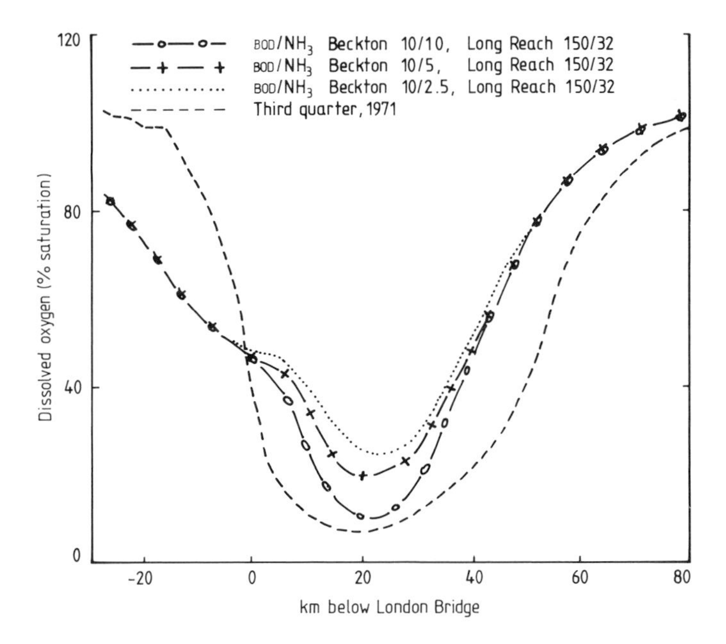

Figure 64. The effect of varying the quality of the effluent from Beckton, maintaining Long Reach at 1974 load.

A similar approach was adopted in the improvement of Long Reach (West Kent) treatment works, situated further downstream. In thise case the model was run using the effluent qualities set out in the table below, for third quarter minimum flow conditions. The results are given in figure 65.

Here it was considered that the difference in river quality produced by effluents (ii) and (iii) was not sufficient to justify the expenditure on a plant to reduce the ammonia nitrogen from 30 to 5 mg/l, especially as the EQO for the Tideway would be achieved by effluent of quality (i). The works was therefore designed to produce a non-nitrifying effluent.

	BOD (mg/l)	Ammonia nitrogen (mg/l)
(i)	20	30
(ii)	10	10
(iii)	10	5

Similarly, the model was used to control, by planning, the siting of heated effluent discharges so that the thermal pollution budget allowed this EQO to be met. The effect of heated effluents on the Tideway has been shown in figure 52. These conditions for the third quarter can be statistically related to corresponding temperatures in May, which were found to be as high as can be permitted if the smolt migration is not to be jeopardised. Modelling the effects of other proposed directly cooled power stations showed that the EQO could not be met if any were sited upstream of Beckton. If power stations were required above this point, then cooling towers would be necessary rather than permitting direct cooling by the use of river water.

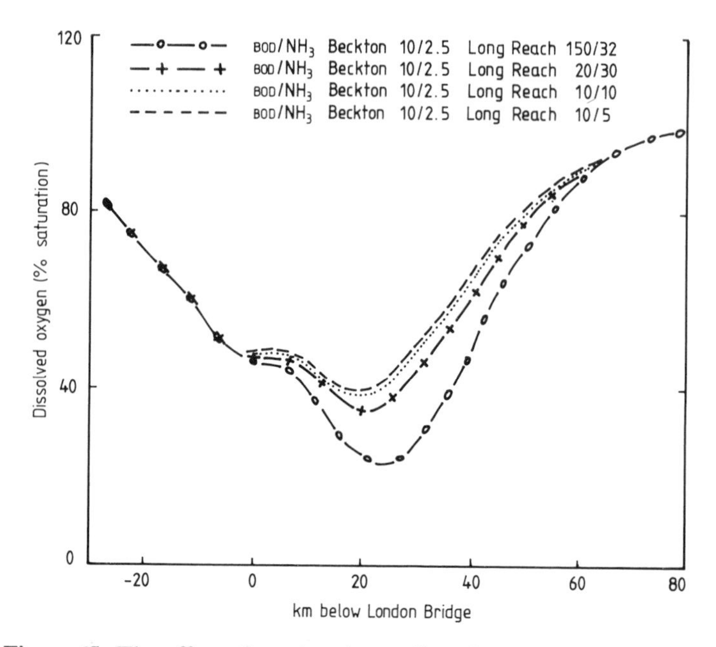

Figure 65. The effect of varying the quality of the effluent from Long Reach, maintaining Beckton at BOD/NH_3: 10/2.5.

(*d*) *Monitoring the Tideway*. The Thames Tideway has been monitored chemically on a regular basis since 1893 from Teddington to the areas where sewage sludge and other spoil, such as dredgings, have been deposited. More recently, biological monitoring has been undertaken to study the changes in fauna and to confirm that the concentrations of toxic substances in fish and other aquatic species do not exceed satisfactory levels. Details of the methods used are given elsewhere.

Upgrading of Sewage Treatment Works

As far as the upper Tideway was concerned, the extensions at Mogden in 1962 dealt satisfactorily with the effluent from that works. From Acton, the sewage had been transferred to Beckton, but during storms there were frequent overflows to the Thames. The final solution was the construction of efficient storm tanks which settle out sludge solids, and retain their capacity of 30 000 m^3 of sewage before a discharge is made to the river. Figure 66 shows four of the six tanks, with the blades of the scrapers which remove sludge from the tank floor for transfer, with the tank contents, to Beckton after the storm has passed. These two improvements, together with upgrading of the sewage works listed in the table below, have led to the present satisfactory condition of the upper Tideway. Figure 67 shows the reconstructed Beddington works, which is fairly typical of a medium-large modern sewage treatment plant.

Figure 66. Storm sewage tanks at Acton.

Figure 67. Beddington sewage treatment works. 1, Storm tanks; 2, primary sedimentation tanks; 3, preliminary treatment plant; 4, diffused air-activated sludge treatment plant; 5, final sedimentation tanks; 6, sludge digestion (under construction), and sludge thickening tanks.

Figure 68. Beckton works, 1960.

Works	Receiving stream
Kew	Thames Tideway
Worcester Park	Beverley Brook
Beddington	River Wandle
Wandle Valley†	River Wandle
Wimbledon†	River Wandle

† Flow transferred to other works.

The Pippard Committee had recommended improvements to Beckton, Crossness and West Kent (Long Reach) to ameliorate the condition of middle and lower reaches of the river, and these were carried out over the following 20 years.

Beckton. Beckton is a very large sewage treatment works, being the largest in Europe, and produces a flow of effluent as great as that of the River Medway, the largest tributary of the River Thames. It remained virtually unchanged from Bazalgette's days until the construction of a rather unusual activated sludge treatment plant. The first unit was built in 1932, followed by others three years later.

The plant consisted of two-tier channels through which the mixed liquor was circulated by Hartley-type paddles, no aeration other than surface aeration being provided. The activated sludge, after being separated in the final sedimentation tanks, was re-aerated in diffused air channels before being recycled. In the post-war period the sanction of the then Minister of Health, Aneurin Bevan was obtained (after his visit to the works) for a reconstruction programme and by 1959 a £7.5 million extension had been completed. This consisted of detritus removal, screen house, primary sedimentation tanks (probably the most efficient of any now operating), a diffused air activated sludge plant of conventional modern design, sludge digestion and a power house using gas-fired turbines. The works, which was capable of giving full biological treatment to 50% of the flow, is shown in figure 68. The extensions were officially opened by the Duke of Edinburgh in 1959.

There was an improvement in the river quality, but even when other works had been extended, it was still necessary for an effluent of a much higher standard to be discharged from Beckton. In 1967, therefore, the Greater London Council, which had replaced the London County Council, authorised a plan to design and construct a treatment plant capable of achieving the effluent quality described in figure 64. The works was designed to treat 1.14×10^6 m^3/day and to be capable of accepting a flow of 2.73×10^6 m^3/day, at a cost of £21 million (1967 prices) (Dainty *et al* 1972, Drake *et al* 1977).

Figure 69 shows the completed Beckton works in 1967. Eight further primary sedimentation tanks were built, each of 11 560 m^3 capacity,

Figure 69. *Top*, Beckton sewage treatment works after the 1967 reconstruction; *bottom*, some of the 48 final sedimentation tanks.

eight diffused air-activated sludge tanks of the 'single pass' type, each of 27 300 m^3 capacity, and 48 further final sedimentation tanks, each of 3500 m^3 capacity, to supplement the 24 already installed. In addition, the sludge digestion plant was extended to provide 32 additional tanks, each of 4750 m^3 capacity, capable of providing digestion treatment of the whole of the sludge. The plan of the works is shown in figure 70.

The completion of the new works at Beckton in 1974 came almost at the

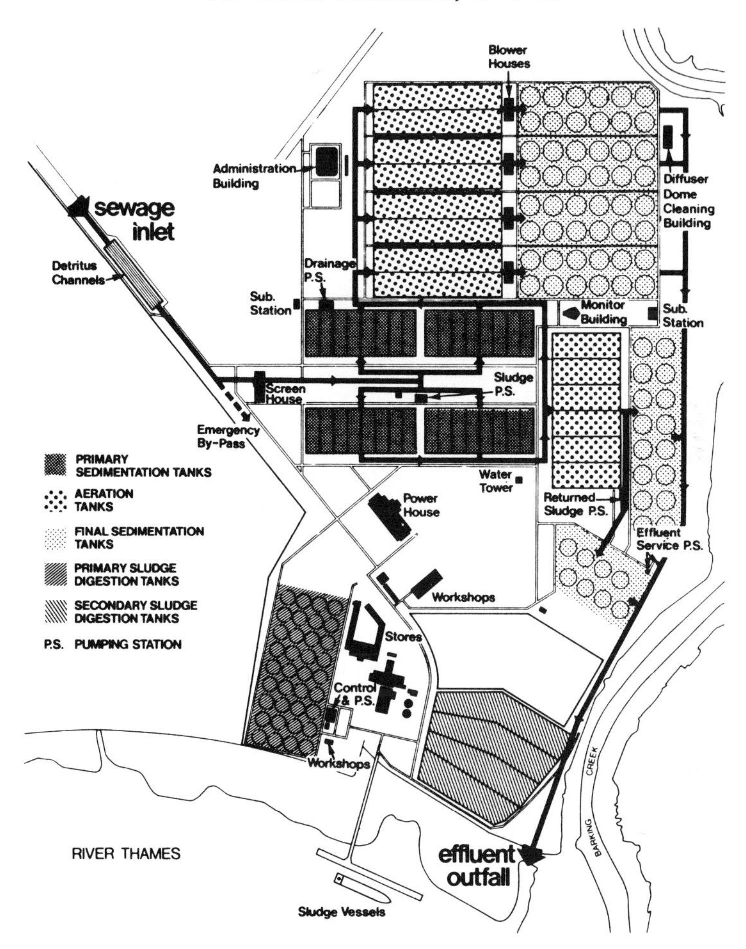

Figure 70. Plan of Beckton sewage treatment works.

end of the reconstruction programme for the riverside works as a whole, and a dramatic change was noticed in the river. The expectations of meeting the EQO, and satisfying the pollution budget were fulfilled. Almost immediately afterwards, the river quality improved sufficiently to allow passage of migratory fish, and in the autumn of 1974 a live salmon was taken from the Thames at West Thurrock, the first since 1833 (figure 82). It is appropriate that this salmon should be held by Mr Peter Black, twice chairman of the Greater London Council and first chairman of Thames Water Authority. As a member he was in the forefront in the motivation of the work.

Crossness. Work began at Crossness in the early 1960s, and the upgraded works was commissioned in 1963. Before that time the works had remained unchanged since its installation by Bazalgette, so that a completely new full treatment plant was designed. Figure 71 shows the plan and figure 72 the whole works. Crossness differs from Beckton in that it is in addition a pumping station, the sewage having to be raised from the low- to the high-level sewers. The works also differs from Beckton in that the activated sludge treatment is based upon 'mechanical aeration', the plant being the largest of its kind in Europe. Instead of the introduction of oxygen by means of fine bubbles of air, as in the diffused air process, rotating cones in the surface of the mixed liquor throw the liquid in fine droplets into the air where, in falling back to the tank, they absorb oxygen at the air–liquid interface. There

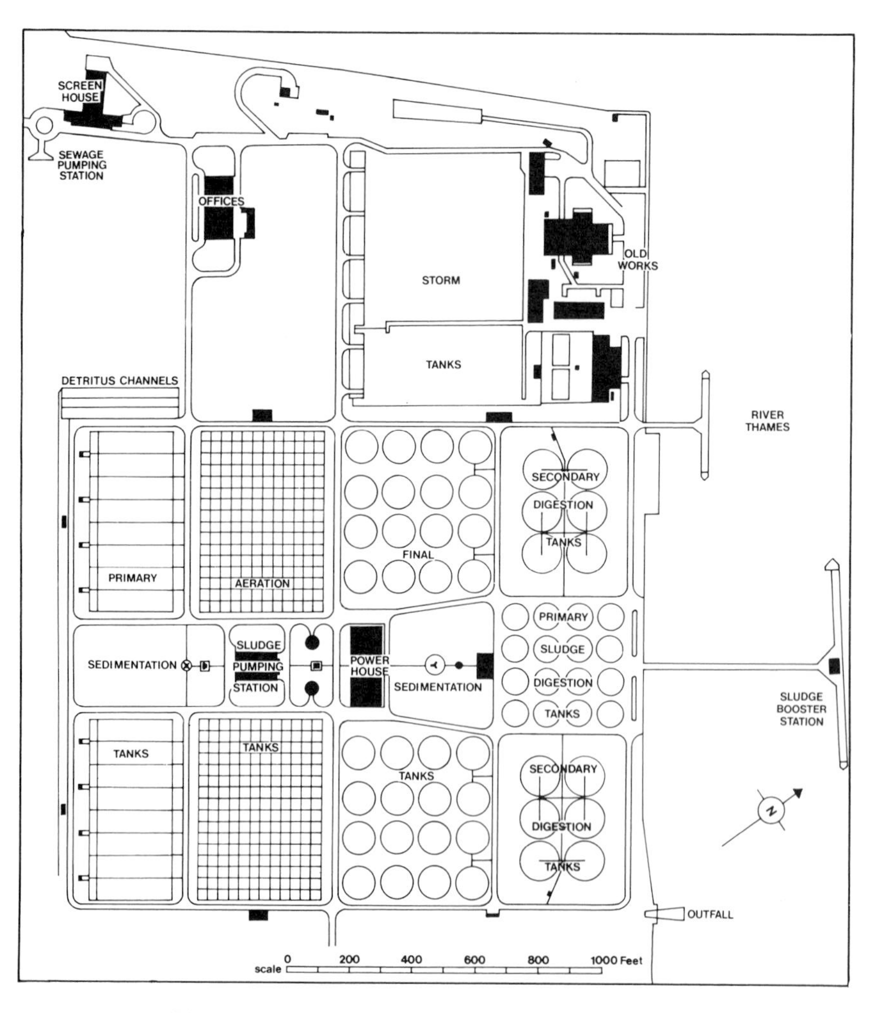

Figure 71. Plan of Crossness sewage treatment works.

Figure 72. Crossness sewage treatment works after reconstruction in 1963.

are 384 cones of about 2 m in diameter driven by electric motors and rotating at 47 rpm. The cost of the works was £9 million at 1963 prices. It was originally designed to treat 0.32×10^6 m^3/day of sewage to the 'Royal Commission' Standard (BOD 20 mg/l, suspended solids 30 mg/l) with no removal of ammonia. However, the plant soon had to cope with 0.45×10^6 m^3/day, and it became clear that a better effluent was required, from which most of the ammonia had been oxidised. The improved effluent quality from the greater flow was achieved by increasing the speed of the aeration cones from 35 rpm as installed, to the present 47 rpm, incorporating at the same time a more efficient design of cone. Crossness was built after the first stage of Beckton improvements, and almost immediately after the works began to produce a nitrifying effluent in 1964, sulphide disappeared from the Tideway. No sample containing sulphide has been found since then, confirming the prediction of the Water Pollution Research Laboratory that nitrate would provide a safeguard against sulphide formation in the river.

Riverside. A different problem arose at Riverside, a medium-sized works having a total flow of 94 000 m^3/day, in that it was receiving about 16 300 m^3/day of trade effluent from a number of factories, one of which discharged compounds which inhibited the oxidation of ammonia in the activated sludge process. If this effluent had been treated in combination with the whole flow to the works, no nitrification (i.e. oxidation of ammonia), would have been possible for any of the effluent. It was therefore agreed to keep this effluent apart from the remainder, and to treat it separately in an activated sludge process which would oxidise only carbon compounds, full treatment being given to the remainder of the flow. The factories discharging these particular effluents, instead of paying normal trade effluent charges based upon the strength and volume of effluent, actually contributed to the construction of the plant itself. The Riverside works is shown in figure 73 and a plan in figure 74. This works uses circular primary sedimentation tanks and mechanical aeration in the activated sludge plant.

Figure 73. Riverside sewage treatment works.

Other works. Reference has already been made to the improvements to Deephams works in 1963; although this discharges to the River Lee, the improvement in that river helped in the upgrading of the Tideway. The last major works to be upgraded was Long Reach, which was completed in 1979. This works provides diffused-air activated sludge treatment and treats the effluent to a non-nitrifying stage. The other works in the seaward part of the Thames estuary have much more relaxed standards because of the greater dilution afforded.

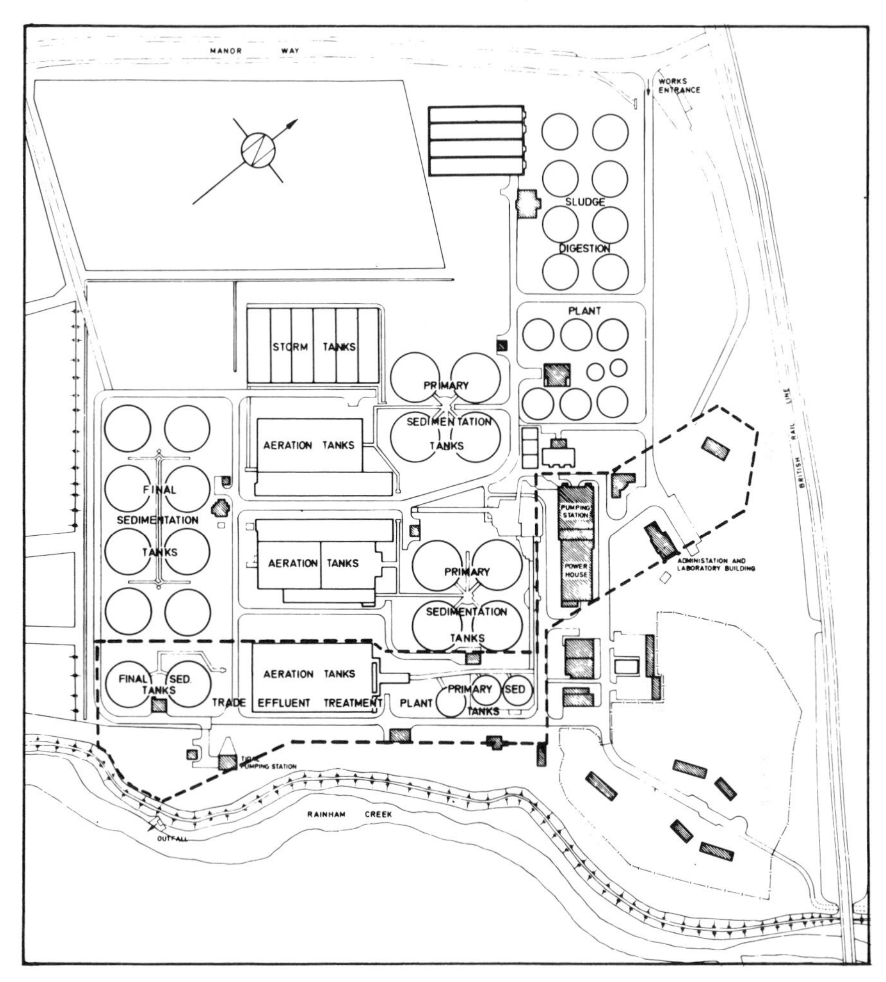

Figure 74. Plan of Riverside sewage treatment works. The broken line represents the trade effluent plant boundary.

Reduction of Polluting Loads

The reduction of polluting loads achieved at individual works, and in the total load of all effluents from the four major works is shown in figure 75. The works extensions and upgrading resulted in a reduction of about 90% in the pollution load between 1955 and 1980†.

† Effluent qualities were actually better than predicted, probably due in a small extent to a reduction in the sewage flow to Beckton.

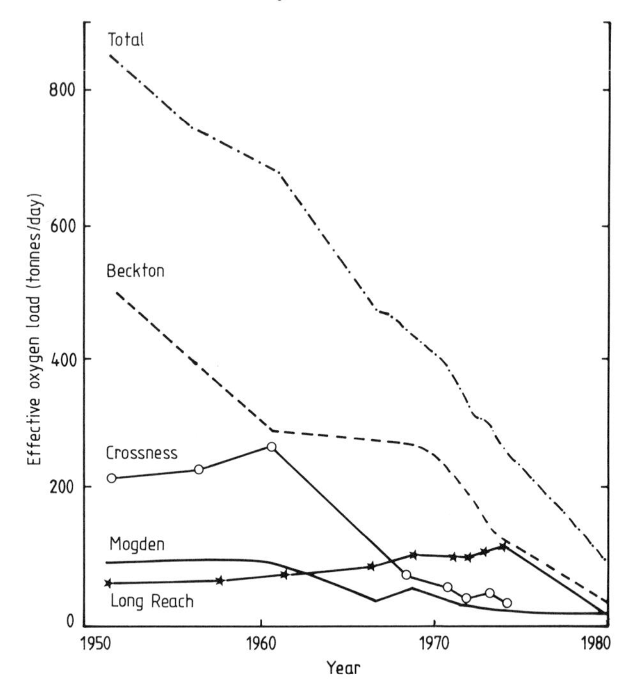

Figure 75. Effective oxygen load (tonnes/day) from the four major sewage treatment works on the Tideway.

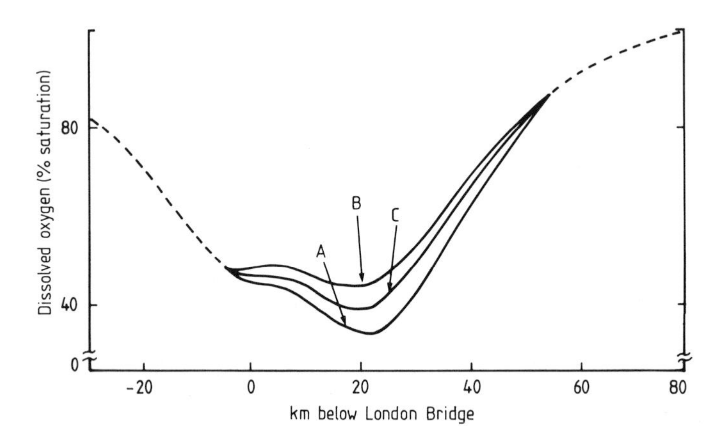

Figure 76. The effect of improving Riverside and Long Reach effluents, third quarter minimum flow conditions. A, 1980 (Long Reach BOD/NH_3 20/30; Riverside 15/36). B, Long Reach 20/30; Riverside 10/21. C, conditions possible with further improvements to secondary treatment. Long Reach 10/2.5.

It has already been indicated that the programme of works improvement was directed towards achieving the EQO required for the river. It is of interest, however, to use the mathematical model to predict what further improvements could have been achieved if the effluents from the major works were all treated to a highly nitrified standard. It appears that the improvement would be as shown in figure 76; the minimum level of percentage saturation with dissolved oxygen would be raised by 10%.

It is probably true to say that the cooperation between industry and the pollution control authorities in Great Britain is the best in the world. It is the declared policy of at least some water authorities not to use their powers of prosecution in cases of illegal pollution, unless these are the result of extreme negligence, wilful actions or repeated offences. This ensures that in the event of accidental spillages, the offender will usually notify the water authority in good time and remedial measures will be speedily effected. This spirit of cooperation has enabled the authorities responsible for management of the Tideway to bring pollution arising from trade discharges to the river to a satisfactory quality. By the time of publication of the Pippard Report in 1961, the load from industrial discharges had been reduced by 60–70% and this trend continued over the next 20 years.

Figure 77. Mechanical aerator installed to increase the oxygen content of the Thames. The ship in the background is the *Bexley*, one of the sludge vessels which transports sludge to the Barrow Deep.

An interesting example of this cooperation arose in connection with the effluent from a paper mill which discharged into the Tideway. It was agreed in 1968 that the company could make the discharge, on condition that they undertook to provide by aeration, dissolved oxygen equivalent to the BOD load of their discharge. The company supplied and maintained two mechanical aerators of the Simplex type which were capable of introducing about 10 tonnes of oxygen per day to the river. Such an arrangement had considerable advantages. The point in the river where the effluent was discharged was not that having the lowest oxygen content, the aerator was therefore installed elsewhere on the river where the dissolved oxygen levels were at a minimum and where the efficiency of oxygen transfer transfer would be greatest. The first of these aerators is shown in figure 77 (the ship in the background is the *Bexley*, one of the sludge vessels which transport sludge to the Barrow Deep).

Improvements in Water Quality

The quality of a river at any point can be judged by its concentration of dissolved oxygen. The levels of oxygen-consuming substances, such as BOD load and ammonia, only allow one to assess possible future quality in terms of the reduction in dissolved oxygen. The levels of BOD from 1920 to 1980 (third quarters) are shown in figure 78(*a*) for the whole Tideway, and in figure 78(*b*) for the river off the metropolitan outfalls. The steady decrease reflects the reduction in the pollution loads of discharges. Similarly, ammonia levels are shown in figure 79. The carbonaceous loads in sewage are removed fairly easily in the early stages of treatment in plants of relatively small capacity. Ammonia requires a substantial quantity of oxygen for its oxidation:

$$NH_3 + 2O_2 \rightarrow HNO_3 + H_2O$$

or, by weight,

$$1\,\text{kg}\,NH_3 + 3.76\,\text{kg}\,O_2 \rightarrow HNO_3 + H_2O,$$

and takes longer, so that larger and more expensive treatment plants are necessary. Consequently, during the early stages of improving the Tideway, ammonia represented a very large demand on the oxygen resources of the river.

The value of nitrate as a reserve of oxygen when dissolved oxygen concentrations are low has already been discussed. As the improved sewage works gave a greater degree of nitrification to the effluents, nitrate levels in the Tideway increased (figure 80). Some denitrification (i.e. decomposition to elementary nitrogen) occurs even in rivers with overall high dissolved oxygen levels, especially in the vicinity of the river bed where anaerobic conditions prevail a few millimetres below the surface. This is particularly noticeable when temperatures are high.

The levels of dissolved oxygen in the Tideway have increased considerably since 1950. The decrease in the length of river with less than 5% dissolved oxygen (i.e. where nitrate reserves are used) has been shown in figure 36. The actual 'sag curves' for the third quarter average conditions from 1950–79 are shown in figure 81. Also included in the latter figure are the average rates of upland freshwater flow over Teddington Weir, which influences both

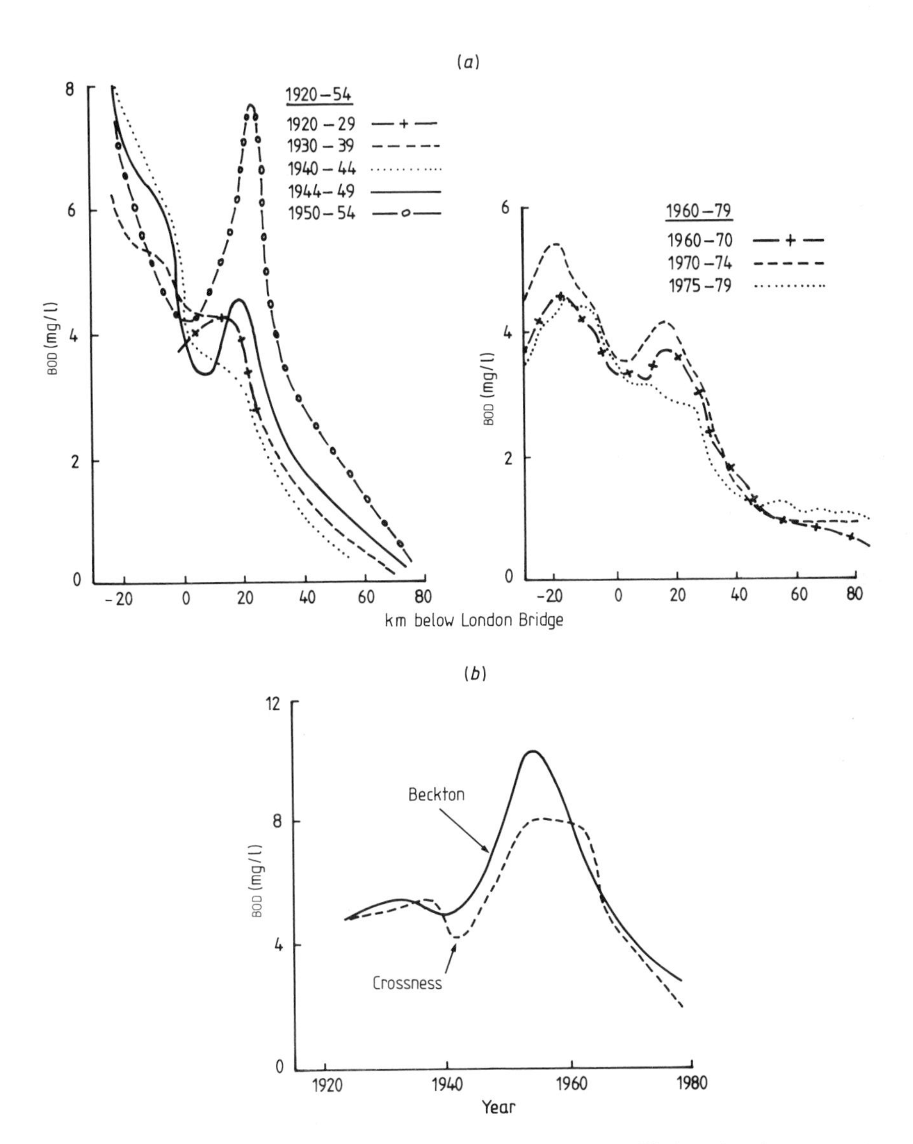

Figure 78. (*a*) Levels of BOD in the Tideway, 1920–80. (*b*) Levels of BOD in the Tideway off the metropolitan outfalls, 1920–80.

the position in the river of the minimum of the sag, and also the dissolved oxygen concentration of the minimum. It is clear that the estuary is now meeting its EQO (not less than 30% saturation with dissolved oxygen) required by the criteria of the Royal Commission on Environmental Pollution.

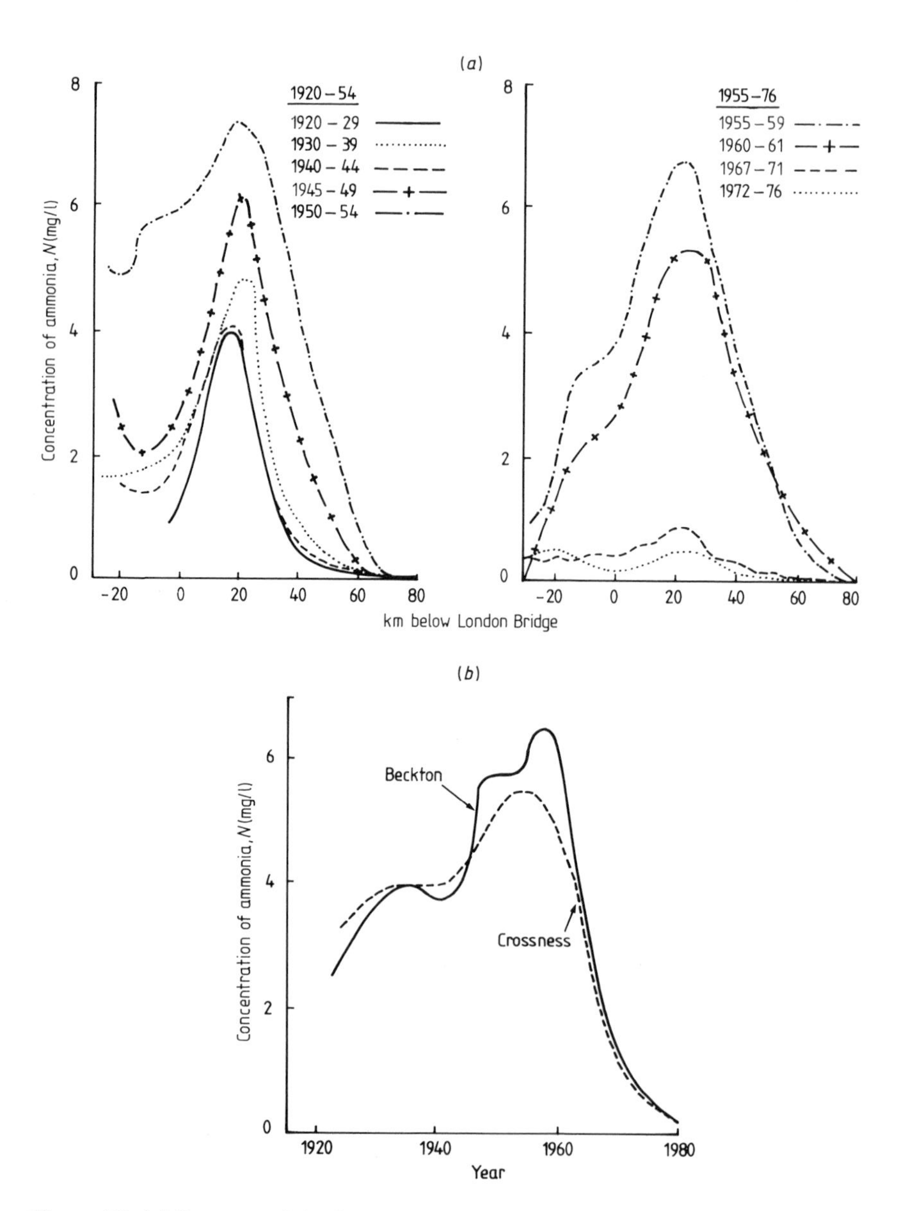

Figure 79. (*a*) Concentration of ammonia in the Tideway, 1920–76. (*b*) Concentration of ammonia in the Tideway off the metropolitan outfalls, 1920–80.

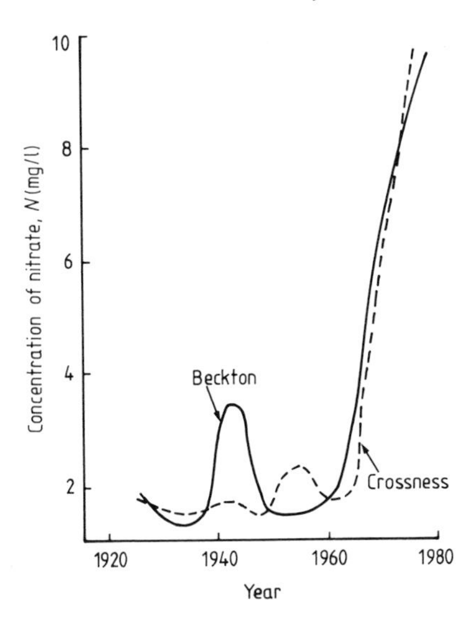

Figure 80. Concentration of nitrate in the Tideway off the metropolitan outfalls, 1920–80.

Royal Commission EQO *Criteria*

Lord Ashby's Royal Commission had suggested the following as criteria for the estuarial conditions required by the public: (1) the ability to support the passage of migratory fish at all states of the tide, and (2) the ability to support on the mud bottom the fauna essential for sustaining sea fisheries. It is now perhaps appropriate to consider whether these conditions have been achieved.

Migratory fish. It is shown in figure 81 that the EQO in respect of dissolved oxygen has been satisfied, and that controlling the siting of directly cooled power stations would ensure that the EQO for temperature would be met. Moreover, the Thames Migratory Fish Committee considered the levels of toxic substances to be satisfactory. It is, however, about 150 years since there were regular runs of salmon, and conditions in the river have changed in other respects. The Thames Migratory Fish Committee indicated physical problems of weirs and similar obstructions to upriver migration, which have developed in the interval. Also, some biologists consider that salmon return to their natal streams by following a trail of low concentrations of pheramones, which are complicated chemical substances specific to a particular hatchery. One cannot but wonder whether the multiplicity of chemicals now made by man may not interfere with this homing instinct, even if present in only minute quantities. Considerable interest has been shown in the possibility of the re-establishment of a salmon fishery since the first was taken alive at West Thurrock in November 1974 (figure 82).

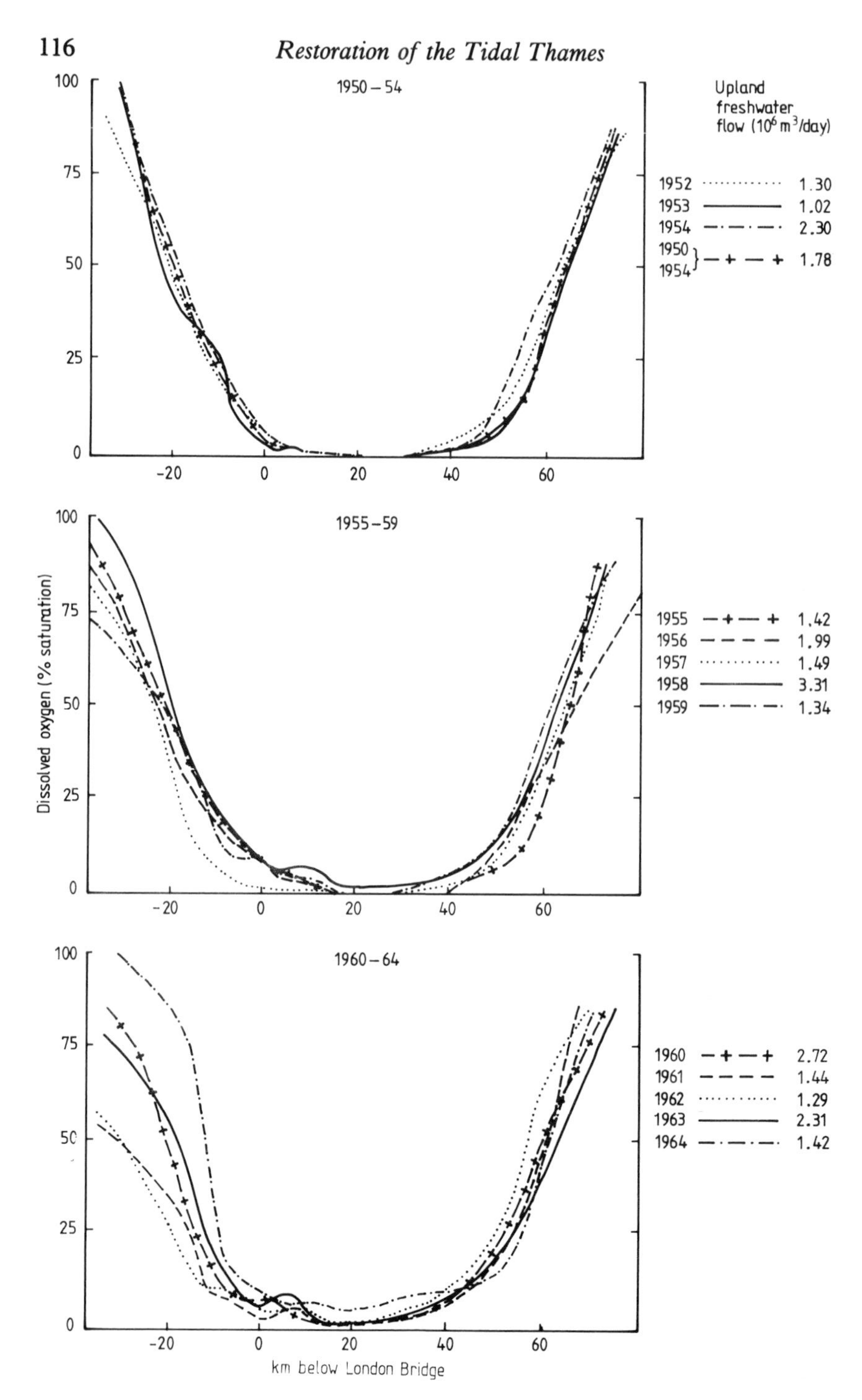

Figure 81. (*a*) Dissolved oxygen concentration, 1950–64 (third quarter averages).

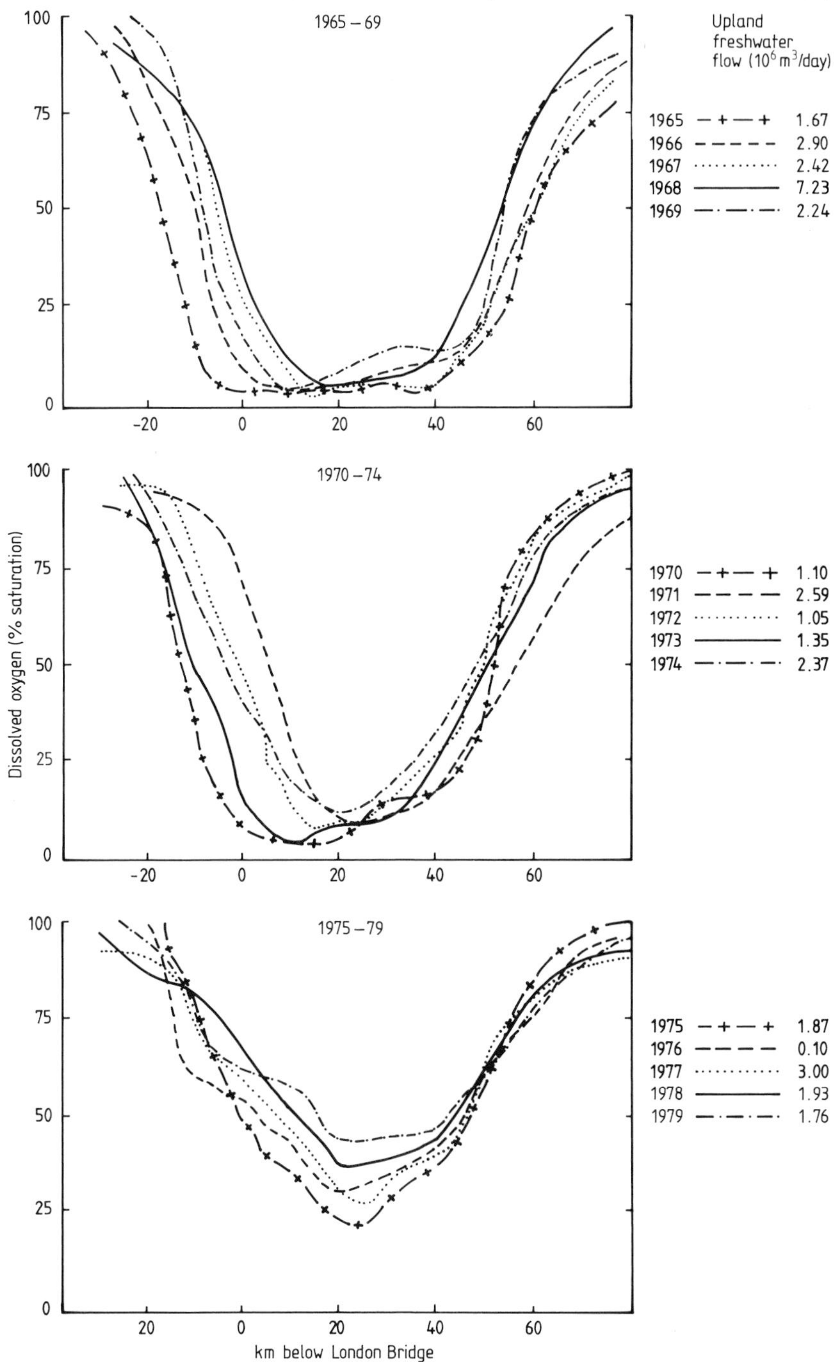

Figure 81. (*b*) Dissolved oxygen concentrations, 1965–79 (third quarter averages).

Experimental restockings using salmon fry from Inverpolly in Scotland have been made since 1974 in the Rivers Pang and Eye in the headwaters of the Thames, in the tributaries to the upper River Lee, and in the River Darent draining to the Tideway. An eminent scientist (Alwyne Wheeler 1979) of the British Museum (Natural History) has given a full account of the migratory fish in the Tideway, and confirms the viability of the salmon taken in 1974, and refers to one fish later found dead in the River Ember (a tributary of the non-tidal Thames) in December 1976, and another seen leaping at Shepperton Weir in the summer of 1978 but eventually found dead in the same weir pool in September 1978. He considers that they may have been vagrants from the salmon stock in the North Sea, and not mature fish from the restocking exercises. He refers, however, to two smolts taken at Brunswick Wharf and West Thurrock power stations, which were presumably from the restocking of the Upper Thames. These catches are a fair indication

Figure 82. Mr Peter Black, Chairman of the Thames Water Authority, holding the first salmon taken live from the Thames since 1833, in November 1974.

that the quality of the river is now good enough to support migratory fish. A future programme for the re-establishment of a Thames salmon fishery is given on p 126.

Sea fisheries. Since the mid-1960s the tidal river has been closely monitored to determine the species, and numbers in the species, of fish in the river. Prior to 1964 the tidal Thames had been devoid of fish life over a stretch of 60 km at least since 1920. There were two well defined stages in the rehabilitation of the inner Tideway by living organisms.

The first stage started in 1964, and was characterised by a gradual but well defined increase in the number of species. The second began in 1976, and was demonstrated by an increased rate of recolonisation by many species. In 1964 the effects of the improvements at Crossness were becoming apparent, in the increased dissolved oxygen levels, and in nitrate from the works effluent, which ensured that after that time the formation of sulphide was prevented. A few fish were found in the neighbourhood of the Dartford Tunnel in that year, the first being the tadpole fish (*Raniceps raninus*), which is a rather rare species, the lampern (*Lampetra fluviatilis*) and the eel (*Anguilla anguilla*), these last two being migratory fish capable of moving upriver through London to the freshwater reaches.

The power stations alongside the Tideway use large quantities of water for cooling. This is first passed through screens to remove coarse floating objects, and therefore any fish present are trapped on the screens. It occurred to Alwyne Wheeler that the collection and enumeration of these fish could afford a useful means of assessing the re-establishment of fish in the Thames. The work started in 1967, and continued as a collaborative exercise with the Greater London Council (Scientific Branch), and finally by the Thames Water Authority, using the power stations listed in table 18. In addition, regular monitoring by trawling and shore surveys (figures 83 and 84) have indicated

Table 18. Power stations used to monitor fish in the Tideway.

Station	Position below London Bridge (km)
Fulham	−11
Lombard Road	−10
Brunswick Wharf	11
Blackwall Point	12
Barking	19
Ford's Works	24
Littlebrook	32
West Thurrock	35

the re-establishment of fauna and provided samples for analyses of fish tissue for monitoring levels of toxic metals and persistent organic substances (table 17).

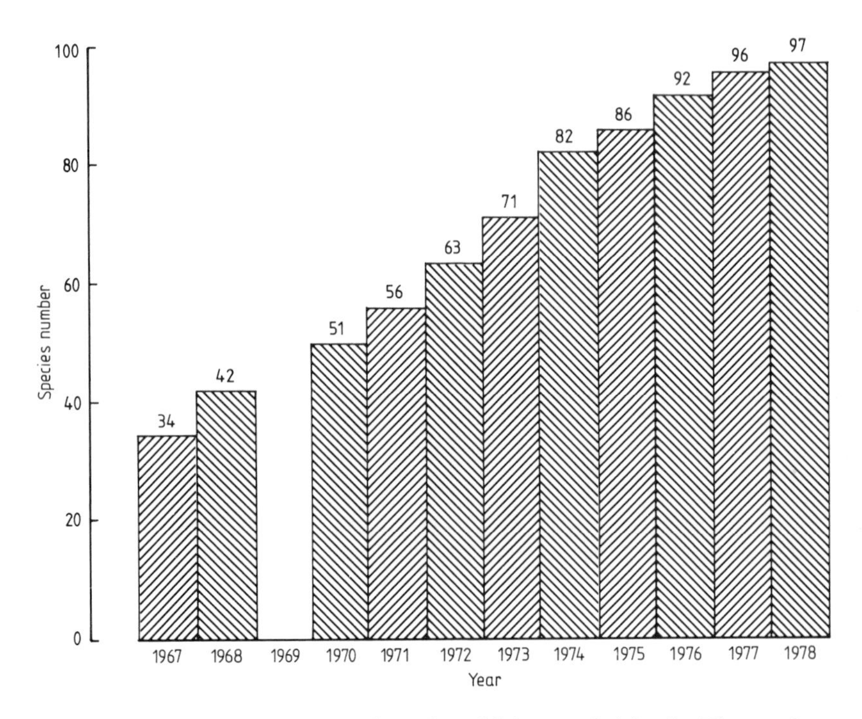

Figure 83. Cumulative number of species of fish recorded in the Thames from Fulham to Gravesend.

Biologists from Thames Water have recently attempted to quantify the degree of ecological improvement, and have identified 1976–77 as the critical period which marked the change from a polluted to a non-polluted regime in the Tideway, based upon various biological parameters. This period coincided with the effect of the improved effluent resulting from commissioning of the final stage of the extension to Beckton treatment works in 1975.

A considerable amount of work has been done in monitoring the return of fish to the Tideway. A great deal of information has been obtained from regular examination of fish taken from the circulating water screens at West Thurrock power station, first by Alwyne Wheeler and later by Michael Andrews. Figure 83 shows the cumulative number of fish species living in the inner Thames since 1964. The increase in numbers of species was almost linear until 1976 (Andrews and Rickard 1980), by which time the majority of the species prevalent in the southern North Sea could be found in the tidal

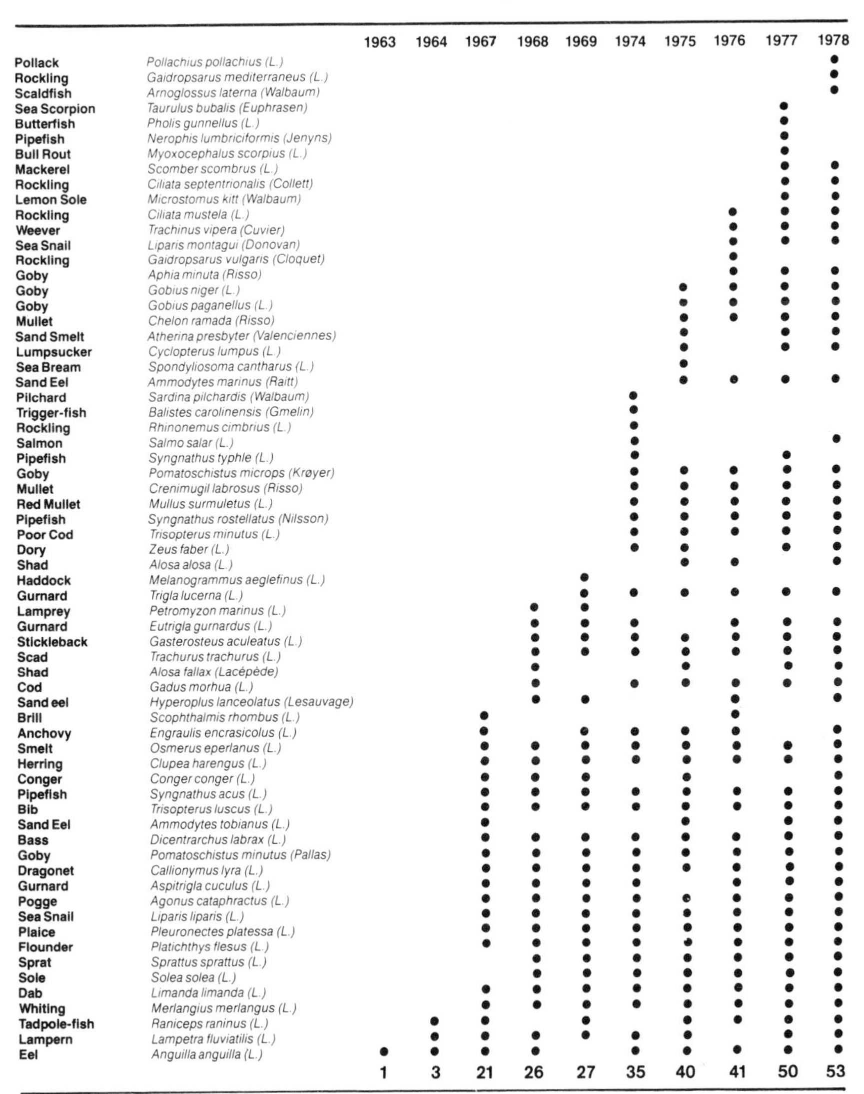

Figure 84. Marine and euryhaline fish recorded at the CEGB intake at West Thurrock power station, 1963–78.

Thames. The results of inspections of fish at West Thurrock power station are shown in figure 86.

The species and numbers of fish tend to be reduced during the warmer months due to offshore feeding habits of gadoids (fish of the cod family such as cod, bib and whiting) and other families. The numbers peak during the

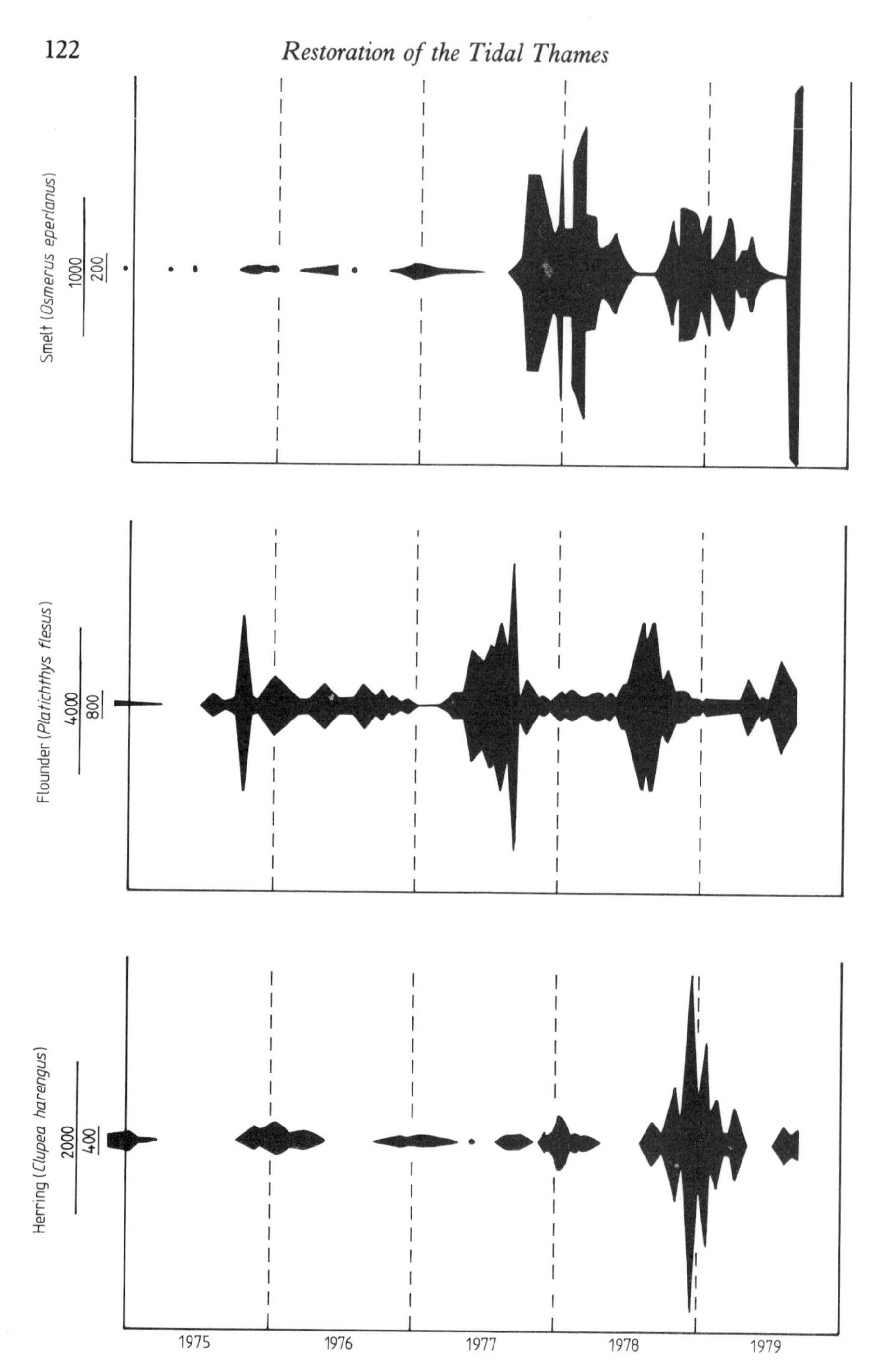

Figure 85. Fish caught at West Thurrock, November 1974 to August 1979 (in numbers per million cubic metres of circulation water.

autumn and winter months when the 0+ and 1+ year groups† enter the estuary. Over the five years 1974–79 the increase in numbers of fish of particular species has been marked in the case of the smelt (*Osmerus eperlanus*), herring (*Clupea harengus*) and flounder (*Platichthys flesus*). The increases in the numbers of fish of these species taken from West Thurrock screens are shown in figure 85. The smelt, which is a true estuarine species, has been found to be breeding in the London reaches as far upriver as Beckton and Crossness, and predominantly 0+ and 1+ year group fish are found at West Thurrock. The increase in numbers of herring is especially encouraging; not only does it vindicate the quality of the Thames water, but it is also significant because the Ministry of Agriculture, Fisheries and Food has banned fishing for this species around much of the British coast to allow depleted stocks to recover.

The numbers of fish in particular species found from year to year varies to such an extent that none can be selected as an 'indicator' species as suggested in Lord Ashby's report. In the winter of 1978–79 the 0+ year group of whiting, for example, was virtually absent, whereas they had been recorded as abundant in previous years. It has been possible, however, to use the data on community structures to obtain a measure of water quality (Gray and Mirza 1979). Using this method, the results for the catches at West Thurrock can be plotted, as shown in figure 86. Where there is slight pollution in the aquatic environment, the data from many communities show a distinct break in the usual straight-line log normal distribution of individuals per species. It can be seen that such pollution was evident in the tidal Thames until 1976, but after that time there was a recovery. The change coincided with improved dissolved oxygen levels in the Tideway.

Macro-invertebrate communities. With changing conditions in the Tideway, Andrews found that the species and numbers of macro-invertebrates have been subject to similar changes. In the early 1970s the fauna in the 'mud-reaches' (2–15 km below London Bridge) was preponderantly of tubificid worms typical of highly polluted waters. The numbers of these worms (*Tubifex tubifex* and *Limnodrilus hoffmeisteri*) were as high as half a million per m^2 of substrate. Even much farther downstream, at Gravesend, where conditions were better, there were few other animals, only ragworms (*Nereis diversicolour*) and acorn barnacles (*Elminius modestus*), apart from the tubificids *Peloscolex benedeni* and *Tubifex costatus*, which are characteristic of brackish water.

The year 1976 saw a coincidence of increased levels of dissolved oxygen in the river and conditions of low upland freshwater flow, which permitted greater incursions of seawater (and hence salinity) into the Tideway. As a

† A fish in its first growing season is placed in group 0 or 0+ while a fish in its second growing season is called a yearling and designated 1+. The age designation by convention changes on 1 January each year.

result, there was a movement of macro-invertebrates upriver towards London. Andrews (1977) listed nearly 50 species, many of which passed through the regions of the major effluent discharges at Beckton and Crossness, where river conditions were poorest. He saw in this evidence that there was no longer a pollution barrier to prevent animal movements in the Tideway. During this time, numbers of tubificids were reduced, and the small amphipod crustaceans *Corophium* spp and *Gammarus* spp, as well as the isopods *Ligia oceanica* and *Sphaeroma rugicauda*, became established in the London reaches. Crabs, shrimps, prawns, sea squirts, jellyfish and sea gooseberries were also plentiful, and the presence of such a varied animal community indicated a healthy biological situation. The stability of the population of the invertebrates was subsequently confirmed after the drought of 1976 when, with salinities at the more usual levels, all the animals remained in the appropriate brackish zones.

Marine algal communities. Tittley and Price (1977) studied the benthic marine algal flora of the tidal Thames between 1968 and 1976, and concluded that the Thames supported a vigorous marine and brackish water algal flora typical of an estuary which had recovered from previous pollution. In the reaches below the major London outfalls the vegetation of the river walls is dominated

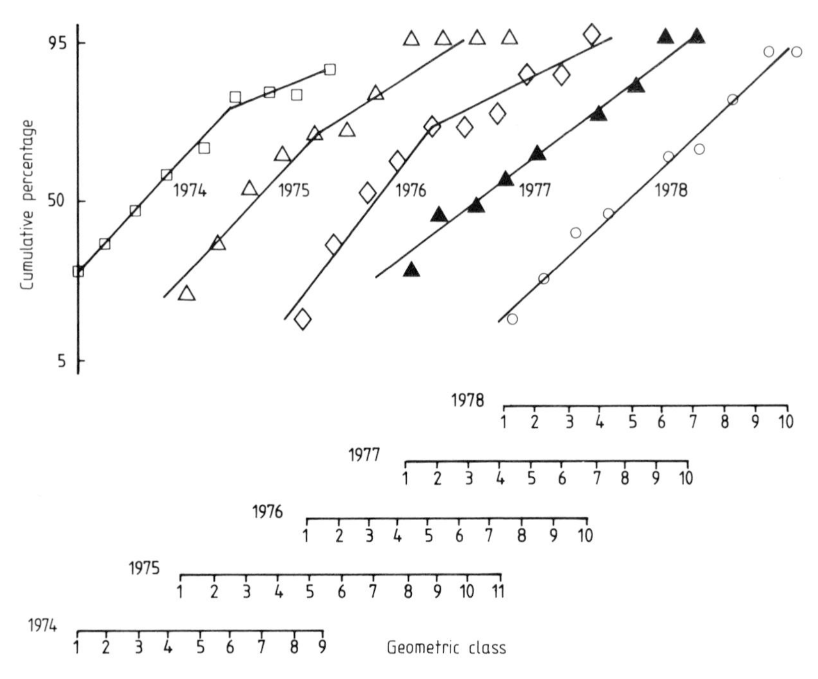

Figure 86. Changes in community structure plots for fish at West Thurrock power station.

by brown algal growths, below a mat of dense green *Blindingia* and *Enteromorpha* spp. Close inspection of the bands of these algae shows the presence of many smaller algae, and in the section upriver of Lower Hope (50 km below London Bridge) the flora is characteristic of that found elsewhere in unpolluted estuaries. In the more seaward parts of the Thames estuary, the rocky shore necessary for these plants is lacking, and the number of their species is lower than in corresponding parts of other estuaries. Where suitable geological structures do exist, however, such as at Whitstable, where hard Eocene clay rocks are exposed at low water, common plants such as *Lamindria saccharina* and *Corallina officinalis* prevail.

Rehabilitation by waterfowl. In the early 1960s there was little birdlife in the area of the inner Thames, but with the recovery of the river in the 1970s, many species of wildfowl and wading birds began to overwinter in these areas. At first, no doubt, the flocks of pochard, tufted duck, shellduck and dunlin were attracted by the abundance of tubificid worms in the reaches in London. Later, the masses of green *Enteromorpha* growing on banks and floating pontoons attracted mallard, mute swan, coot and moorhen. It was estimated that in the cold winters of the mid-1960s up to 10 000 wildfowl and 12 000 waders returned to the inner Thames. This was a great achievement, since any area of Europe used by 10 000 birds or more is designated by the International Waterfowl Research Bureau as a wildfowl refuge of prime international importance.

With the change in the fauna of the Thames from an aquatic one dominated by tubificids, there has been a corresponding change in the waterfowl communities. Whilst large flocks of pochard are usually counted (5800 at Crossness in January 1979), more of the rarer species are now observed. Shovellers feed on the abundant surplus of small crustaceans in the bay at Barking, and lapwings, redshanks and ringed plover are common. The less common greves, red-breasted mergansers and Leach's petrel have also been observed. Unfortunately, despite the greater abundance of fish, the species of birds for which they provide food have not become very numerous. Although such fish-eating wildfowl as the goosander, red-throated divers and smew are found in winter counts, they are not numerous, and are not likely to increase because the Thames is a silty river, and poor underwater visibility prevents the birds from easily catching their prey.

Further Measures to be Taken

Restocking the Thames with Migratory Fish

On page 94 strategies were set out which would be required to attain the standards of control outlined in the Royal Commission Report. The foregoing description of remedial work carried out, and the results achieved, show that these aims have been met and that the river is now of a quality which complies with the recommendations of the Report. Nevertheless, although the quality is thought (in the present state of our knowledge) to be capable of supporting the passage of migratory fish, whether it will do so to the extent that healthy communities of salmon and sea trout can be established, is yet to be proved. This will be seen on the completion of the programme of restocking (Thames Water Authority 1978).

In order to try to foresee the success of this operation it is perhaps pertinent to look at the requirements of migratory fisheries and to study the life cycles of these fish.

After hatching in freshwater streams, the Atlantic salmon (*Salmo salar*) spends up to three years in fresh water before migrating to sea, where its growth is much more rapid. It then returns to fresh water, matures and spawns, a large proportion of the adult fish dying immediately thereafter. The hen (female) digs a redd (nest) in gravel in a suitable stream bed, where the eggs are deposited and fertilised by the cock (male). The eggs incubate over the winter and become alevins (having an attached yolk sac) and then develop into fry about an inch long, which then emerge through the gravel in the spring. The fry develop in their first year into parr and eventually migrate to sea, assuming beforehand a silvery colour characteristic of their growth stage as smolts. Salmon can spend up to three years at sea, sometimes travelling up to 3500 km before returning. Many return to the river from which they came; some even to the stream of their hatching.

The life cycle of the sea trout (*Salmo trutta*) is similar to that of the Atlantic salmon. It hatches into a smolt, then migrates to sea, and subsequently returns to fresh water to spawn. Young sea trout grow faster than salmon and tend to migrate earlier than salmon smolts. They migrate in the spring and return in the same year to fresh water as whitling (or salmon peal) and may actually spawn that year. Another difference is that more adult trout survive after spawning, and may return several times for the purpose. Sea trout do not usually travel so far from their natal streams as do salmon, and spend their time at sea around the coastal area of the country. Compared with salmon, the homing instinct is not so apparent in returning adult sea trout, which may spawn in other than their natal rivers.

There are therefore different problems involved in re-establishing fisheries for salmon and for sea trout. The former will require that the appropriate freshwater stream is restocked with eyed ova, feeding fry, parr or smolts at one year, in order that a return from the sea will be achieved. Because of the less well developed homing instinct of sea trout, these fish will re-establish

Table 19.

Growth stage	Rate of survival to smolt stage (%)	Number required for 10000 smolts
Eyed ova	2.4	400 000
Feeding fry	5	200 000
Parr	10	100 000

Table 20. Rivers suitable for salmon spawning.

	Area of bed available (ha)
Upper Thames area	
River Evenlode	10.0
River Kennet	9.2
River Windrush	8.8
River Coln	6.0
River Enborne	5.2
River Churn	3.0
River Pang	3.0
River Lee area	
River Rib	9.3
River Lee	7.0
River Lee flood relief channel	6.0
River Mimram	3.0
Tideway area	
River Darent	1.0

themselves from stocks from other rivers, provided that suitable environmental conditions exist. Unfortunately, owing to the action of predators, the survival rate of young salmon is not high, so that immense numbers of young fish are required (see table 19). In attempts to restore populations of salmon it is necessary to have suitable gravel-based streams for the hen to construct her redd. Suitable areas are shown in table 20. The total suitable area available in the Thames catchment is 72 hectares, and it is considered that for the production of 10 000 smolts, 19 hectares of gravel river bed would be required.

Since salmon last populated the Thames, however, many installations have been constructed which could obstruct migration. Including Teddington, there are 44 weirs on the freshwater river, and many other obstructions on the tributaries which would be used by migrating fish. Some of these will require modification before salmon will be able to surmount them, either by allowing the salmon sluices through which to pass or by providing suitable weir pools for them to leap the weirs.

The Rehabilitation Programme

To supply the necessary fish and to provide free access for their migration is expensive, and it is proposed to carry out a programme phased over a number of years to establish a satisfactory community in the Thames.

Phase I (duration seven years). Juvenile salmon will be reared in the Thames Water Authority's fish farms and nurseries. Returning migratory fish will be stripped and used to provide ova for rearing young fish. The success of all phases of the operation depends upon this stage, and it will decide whether the establishment of a salmon fishery in the Thames is viable.

Phase II (duration five years). This will of course be subject to the success of the first phase, the rearing programme will continue but, in addition, attempts will be made to encourage natural spawning in the Thames and its tributaries. In phase I fish would not be able to ascend tributaries further upriver than the River Mole confluence, but in the second phase modifications to weirs would allow their passage to Cookham, so that they could ascend to the Rivers Wey and Colne.

Phase III (duration five years). In this phase, the number of fish provided by artificial means of propagation would be reviewed in the light of the natural spawning and rearing. Only the fish most suitable genetically would be used for stripping, so that the best quality stock would develop.

If this restocking programme is successful, then the River Thames will indeed have been rehabilitated.

Control of Storm Sewage Discharges

Reference has already been made to the pollution which arises from storm sewage overflows, but few of the remedial measures undertaken since 1950 have been designed specifically to mitigate their effects. Before the quality of the river was restored the effects of such discharges often went unnoticed, since there were no fish to be killed. There have been two occasions more recently (in 1973 and 1977) when, as a result of severe storms, there were substantial fish kills in the Tideway. The 1973 storm resulted in the loss of about 1000 coarse fish, and on the second occasion, although the improved conditions still permitted coarse fish to survive, the quality was unsatisfactory for the sensitive smelt (*Osmerus eperlanus*), many of which died.

Whenever there is a sudden heavy storm, particularly after prolonged dry weather, the river is likely to be very seriously affected. The upland flow of fresh water is too low to provide a high degree of dilution, and the run-off from the impermeable urban area is heavy, so that the increased flow in the combined sewers makes it necessary to transfer large volumes of storm sewage to the river. Because the major sewage works at Beckton and Crossness will be subject to greatly increased loads, their effluent quality will deteriorate in the short term. Thus, since storm sewage outlets (figure 26) are situated upstream of the major outfalls, fish could be trapped between two masses of water low in dissolved oxygen, especially with a making flood tide. The two oxygen sags then converge and coalesce, leaving no escape for fish which may then be killed.

Recent Significant Storms

June 1973: Rain started in London on 19 June 1973 and continued during the following day, the resulting average precipitation being 60 mm over the two days. As a result 6.4×10^6 m^3/day of storm sewage were discharged to the river, and a further 2.5×10^6 m^3/day from surface water sewers to the Thames tributaries. After the previous dry period, the latter discharges were probably contaminated with accumulated polluting materials washed off the streets. The resulting polluting load was estimated as being 280 tonnes of effective oxygen load (EOL) (see Appendix I), which reduced the dissolved oxygen levels in the Tideway to below 5% saturation over a 25 km stretch, including locations where the level had fallen almost to zero. Figure 87 shows the river condition prior to the storm, the development of two oxygen sags on the second day of the storm, and finally the coalescence of the oxygen sags seven days later. On this occasion there was a fish mortality of about 1000, mainly of flounders.

August 1977: On 16 August 1977 a storm started when the upland flow was a little higher than in the previous case, and resulted in an average precipitation of 42 mm over the London area. The result was a substantial fish kill in the area of Woolwich Reach, 20 km below London Bridge. The fish were smelt, which are known to be intolerant of dissolved oxygen concentrations of less than 30% saturation; conditions were observed to be quite acceptable to coarse fish seen in the reach. The progress of the storm was similar to that

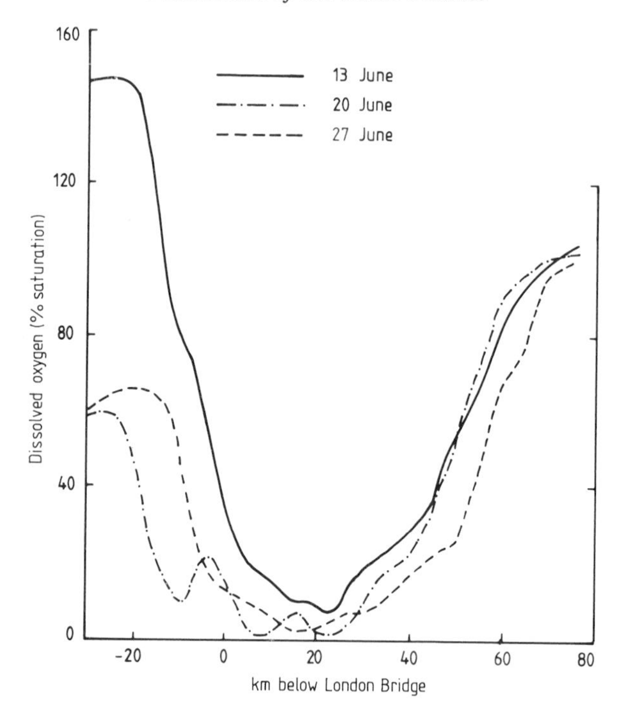

Figure 87. Dissolved oxygen curves for the storm of June 1973.

of 1973. Figure 88 shows the conditions immediately before the storm, and as it developed after the first day. Since this storm occurred at a time when the condition of the river was considered to be almost at the maximum quality likely to be achieved, it was studied in considerable detail.

One difficulty which besets this kind of study is that whereas those storm discharges which are pumped to the river are analysed and their flow recorded, it is not possible to obtain this information for discharges made by gravity. A method has been suggested for sampling and measuring flows in gravity storm sewers, consisting of a switched system which is brought into operation when the level rises in the sewer, and this initiates a dosing system whereby lithium chloride solution of known concentration is injected upstream of an automatic sampler, which is also brought into operation. When samples are collected they provide, in addition to that taken for normal sanitary analysis for the pollution content, a sample of diluted lithium chloride, from the concentration of which the flow can be calculated. These systems are, however, very expensive, and difficulties arise in ensuring that a true sample is being taken, especially when river water backs up into the sewer at high spring tides. At present, therefore, they are not extensively used, but a very rough approximation of the pollution load can be computed from that of the pumped discharges, the ratio of pumped to gravity loads being 100:15.

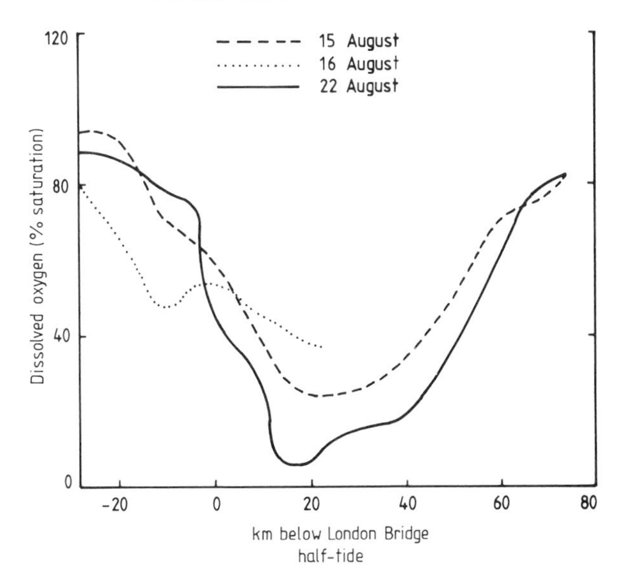

Figure 88. Dissolved oxygen curves for the storm of August 1977.

The magnitude of the loads of storm sewage on 16 August 1977 is given in table 21. During the storm the pollution loads from Beckton and Crossness also increased to 105 tonnes/day due to the increased volume and poorer quality of effluents resulting from the greatly increased flows of sewage received (table 22). In addition, the load imposed from the tributaries rose to an extent which was not accurately known, but was thought to be considerable. The discharge from the storm sewage overflow can be seen to have been a total of 394 tonnes/day of EOL (table 21), compared with a combined flow of 105 tonnes/day from Beckton and Crossness.

Alleviating the Effects of Storms

The strategies for obviating the effects of storm sewage pollution must be either to reduce the pollution load entering the river, or to compensate for this load by increasing the dissolved oxygen levels.

Reduction of the pollution load. Two methods are available for this purpose:

(*a*) The dimensions of the intercepting sewers could be increased and the treatment capacity at the receiving works extended to deal with the additional flow. This would be prohibitively expensive, as well as uneconomic, since these extensions would be unused for much of the time. In 1977 the cost was estimated to be as high as £70 million.

(*b*) Storm tanks and similar devices (Prus-Chacinska 1976, Field *et al* 1977) have been found to be effective in reducing the pollution load by removing the solid matter prior to discharge of the sewage. In areas adjacent to the

Table 21(*a*). Storm sewage discharges to the tidal Thames below London Bridge.

Discharge	Map ref. no.‡	Method of discharge	Position below London Bridge (km)	Flow (m^3/day)	EOL† (tonnes/day)	Oxygen demand of discharge* (tonnes/day)					
						1	2	3	4	5	6
Gascoigne Road	1	Pumped	19.0	38 100	4	0.88	0.66	0.50	0.39	0.29	0.23
Folkestone Road	2	Pumped	19.0	125 300	13	2.89	2.18	1.65	1.26	0.97	0.74
North Woolwich	3	Pumped	16.2	92 000	9	2.12	1.60	1.22	0.92	0.71	0.54
Charlton	4	Gravity	13.1	85 800	9	1.98	1.49	1.13	0.87	0.65	0.51
Canning Town	5	Pumped	11.1	49 100	5	1.13	0.86	0.64	0.50	0.38	0.29
Abbey Mills	6	Pumped	11.1	803 900	80	18.53	14.00	10.60	8.09	6.19	4.75
Isle of Dogs	7	Pumped	9.9	113 000	11	2.60	1.97	1.49	1.14	0.87	0.67
Deptford	8	Pumped	7.1	585 600	59	13.50	10.19	7.73	5.89	4.51	3.46
Deptford Green	9	Gravity	6.8	85 800	9	1.98	1.49	1.13	0.87	0.65	0.51
Earl	10	Pumped	5.4	61 200	6	1.41	1.07	0.80	0.62	0.47	0.36
Holloway	11	Gravity	3.6	21 500	2	0.50	0.37	0.28	0.22	0.17	0.12
North-eastern	12	Gravity	3.0	42 900	4	0.99	0.75	0.56	0.43	0.33	0.26
Shad	13	Pumped	1.1	185 500	19	4.28	3.22	2.45	1.87	1.43	1.09
Total				2 289 700	230	52.79	39.85	30.18	23.07	17.62	13.53

Table 21(*b*). Storm sewage discharges to the tidal Thames above London Bridge.

Discharge	Map ref. no.‡	Method of discharge	Position above London Bridge (km)	Flow (m^3/day)	EOL† (tonnes/day)	Oxygen demand of discharge* (tonnes/day)					
						1	2	3	4	5	6
Fleet	14	Gravity	1.0	42 900	4	0.99	0.75	0.56	0.43	0.33	0.26
Brixton	15	Gravity	4.3	42 900	4	0.99	0.75	0.56	0.43	0.33	0.26
Clapham	16	Gravity	4.4	42 900	4	0.99	0.75	0.56	0.43	0.33	0.26
Heathwall	17	Pumped	5.0	139 500	14	3.22	2.42	1.85	1.40	1.07	0.83
South-western	18	Gravity	5.0	21 500	2	0.50	0.37	0.28	0.22	0.17	0.12
Western	19	Pumped	6.2	383 800	38	8.85	6.67	4.77	4.17	2.95	2.27
Ranelagh	20	Gravity	6.4	21 500	2	0.50	0.37	0.28	0.22	0.17	0.12
Counter Creek§	21	Pumped	8.4	113 300	11	2.61	1.97	1.50	1.14	0.87	0.67
Walham Green§	21	Pumped	8.4	189 800	16	3.68	2.79	2.10	1.61	1.23	0.95
Falconbrook	22	Pumped	9.3	84 600	8	1.95	1.47	1.12	0.85	0.65	0.50
Wandle Valley	23	Gravity	9.6	21 500	2	0.50	0.37	0.28	0.22	0.17	0.12
Hammersmith	24	Pumped	14.9	497 900	50	11.48	8.67	6.57	5.01	3.85	2.94
Acton	25	Gravity	15.7	85 800	9	1.98	1.49	1.13	0.87	0.66	0.50
Total				1 687 900	164	38.24	29.08	21.56	17.00	12.76	9.80

† EOL (effective oxygen load) = $F \times 10^{-6}(1.45B + 4.57N)$ tonnes/day, where F = flow (m^3/day); B = five-day BOD (ATU) (mg/l); and N = total oxidisable nitrogen (mg/l).

‡ See map, figure 26.

§ These two pumping stations share the same outlet.

*Oxygen uptake by storm discharge on first day of storm and subsequent days.

Table 22. Loads discharged from Beckton and Crossness sewage treatment works during the storm of August 1977.

Works	Position below London Bridge (km)	Date August 1977	Flow (10^3 m^3/day)	EOL (tonnes/day)
Beckton	18.3	16/17	1250	61
		17/18	1603	38
		18/19	1093	25
		19/20	1220	23
Crossness	21.8	16/17	824	44
		17/18	803	20
		18/19	602	16
		19/20	645	16

Thames, however, there is not enough land available to build sufficient capacity into these systems to deal with the enormous flows involved, and in addition the sewerage system of London is so complex with cross connections that this would be extremely expensive.

Compensation by the addition of oxygen (Wood *et al* 1978, 1979a). The quantities of oxygen required to satisfy the demands made by storm sewage on the Tideway in storms such as those of 1973 and 1977 are such that they cannot be replenished by the introduction of oxygen in air; 'pure' oxygen† must be used (see Appendix I). Methods of achieving this compensation are as follows:

(i) Pumped storm sewage discharges could be injected with oxygen as they pass through the station. It has been shown (Imhoff and Albrecht 1977) that if water is oxygenated beyond 200% of air saturation, there is likely to be a loss of gas to the atmosphere, and the process will be inefficient. Figure 89 shows the rate at which the pollution load of storm sewage will consume oxygen during its passage in the river. Even if oxygen equivalent to the total load could be injected at the pumping stations, it would be about 10 days before most of the reaction was completed. During that time undoubtedly most of the oxygen would have been wasted by loss to the atmosphere. On the other hand, if the pumped discharges were oxygenated to 200% saturation, this would provide, for those within 15 km upstream of London Bridge, 39 tonnes of oxygen which, even assuming it were completely used, would only just suffice for the first day's uptake (36 tonnes).

(ii) Tributaries of the Thames and effluents from sewage treatment works could be oxygenated, but the oxygen would not necessarily be injected where it is required, i.e. at the lowest point of the upriver oxygen sag.

† 'Pure' oxygen refers to a gas containing 90–95% oxygen, as distinct from air, which contains only 20% oxygen.

(iii) Power station cooling water could be oxygenated, but as the smaller London riverside power stations are being run down, insufficient water is pumped, especially during the summer months, to guarantee a solution.

(iv) The most suitable solution appears to be to inject oxygen directly at the point in the river where the concentration is lowest, using purpose-built vessels equipped with oxygenating equipment. Economic appraisal shows that the most suitable vessel is a 'standard' Thames barge.

Oxygen Injection from Barges

A storm of the magnitude described would require about 30 tonnes of oxygen per day. If containers of liquid oxygen were used as a source of supply, they would not only be excessively bulky, but in a river used for industrial transport, the system could also pose dangers of fire and explosion if the liquid came into contact with organic materials such as hydrocarbons. The preferred method of obtaining oxygen is the so-called 'pressure swing adsorption' process, which generates a gas of very high oxygen content from the air. The process makes use of the fact that when air (containing approximately 80% nitrogen and 20% oxygen), is passed over beds containing pellets of 'molecular sieves', nitrogen is adsorbed (i.e. held by forces of attraction to the surfaces) preferentially to oxygen. After the first pass over the bed, the effluent gas, whilst enriched in oxygen by virtue of the nitrogen held on the bed, still contains too much nitrogen to be acceptable, and it must therefore be recirculated until the oxygen content is satisfactory. Also, after the first passage of air, the bed must be stripped of nitrogen by applying reduced pressure, and the gas exhausted to air. It is therefore usual to have three tanks containing molecular sieves, alternately adsorbing and being stripped

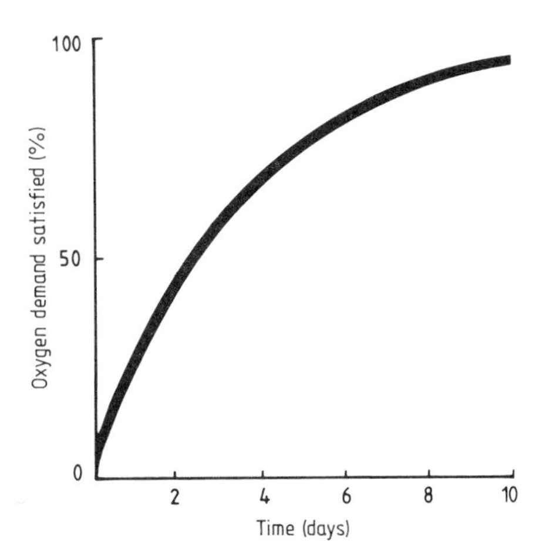

Figure 89. Oxygen demand throughout the August 1977 storm.

of nitrogen, and recirculating the oxygen-enriched gas. Desiccants are mixed with the molecular sieves to absorb water vapour which would otherwise reduce their efficiency considerably.

If a conventional system of diffusing oxygen into the river were used, and the oxygenation limited to what is considered suitable in practice, namely 200% of air saturation, then in order to inject the 30 tonnes of oxygen per day required to satisfy the oxygen demand, about 1.5×10^6 m^3/day of river water would have to be pumped and circulated. This flow, which is equal to the combined average daily flows of Beckton and Crossness, would require at least ten standard barges to house the pumps. There is however a proprietary injection system—'Vitox'†—by means of which 30 tonnes of oxygen can be added in only one-tenth of this volume of water. The system (figure 90), involves pumping water from the river and injecting oxygen in fine bubbles into a high-pressure stream of water. When this is returned to the river through expansion nozzles, the bubbles are shattered on impact to extremely small bubbles which rapidly dissolve and achieve a high degree of oxygen transfer.

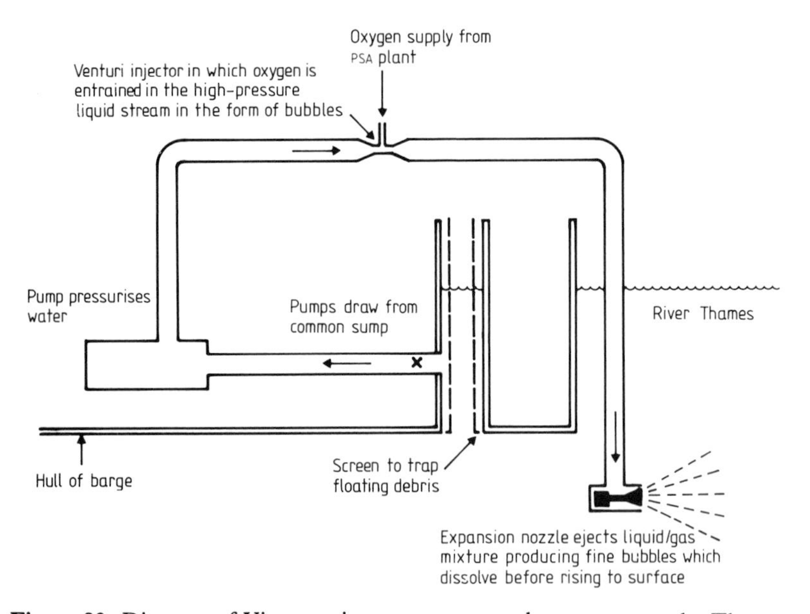

Figure 90. Diagram of Vitox equipment as mounted to oxygenate the Thames.

The Vitox system, together with pressure swing adsorption equipment when mounted on a barge (figure 91), affords a suitable means of oxygen injection at the lowest point in the upstream dissolved oxygen sag, during

† Vitox is a patented process of British Oxygen Co Ltd.

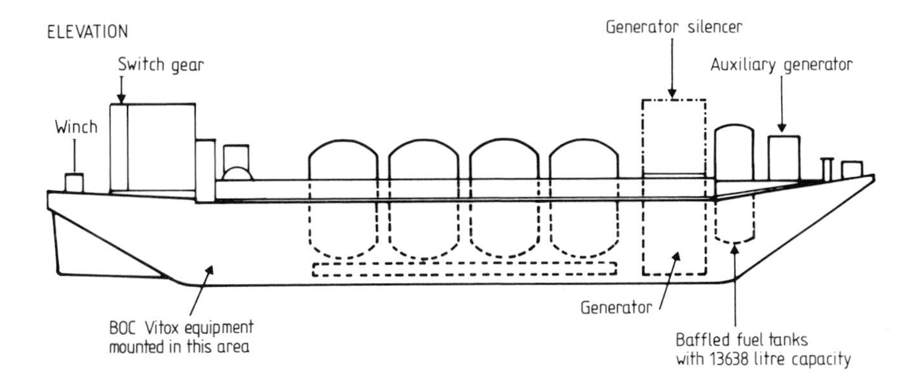

SECTION

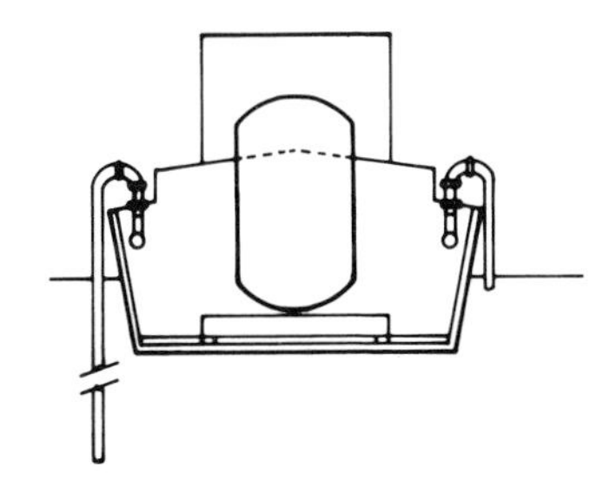

PLAN

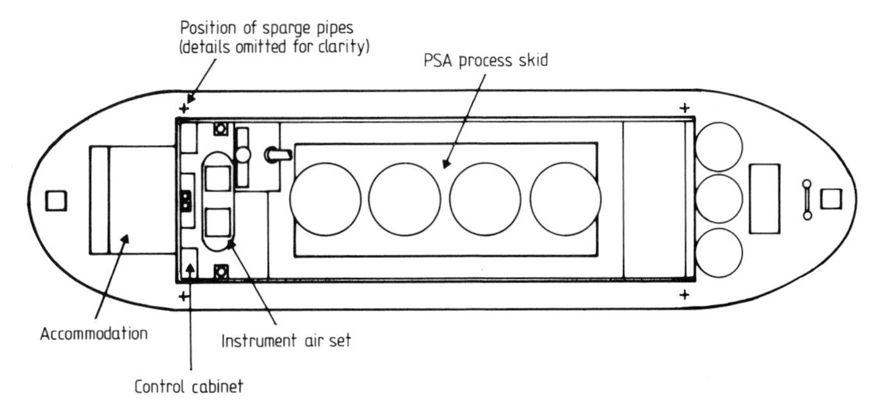

Figure 91. Barge for emergency oxygenation of the Thames. (Length 26.7 m, beam 6.8 m, height 2.1 m; weight 52 tonnes unladen, 160 tonnes laden, draught laden 1.4 m.)

the persistence of the effects of storm sewage discharge. Such a system has been designed for this purpose to be used on the Thames.

Ideally, if it were possible to construct a satisfactory time-dependent mathematical model which would indicate effects of discharges over short periods of time, this could be used to predict the minimum point of the oxygen sag at any time after the start of a discharge. So far, it has not been possible to devise a sufficiently accurate model. However, reference to table 21 shows that discharges can be conveniently divided into two groups. Those below London Bridge will contribute to the downstream sag (i.e. that which arises mainly from the effects of the metropolitan sewage works' effluents and which occurs even in dry weather), and any remedial oxygen injection could be made efficiently by injection at Beckton works. Those storm discharges between London Bridge and a point 15 km above it contribute to the upper sag, and it is this section which must be treated by the oxygen barge. The barge would be moored above London Bridge and brought into operation when the rainfall pattern indicates its necessity. It would be navigated upstream, determining by an electrode the level of dissolved oxygen, until the point is reached where oxygen injection should start. Thereafter the river would be monitored by a launch to determine the pattern of dissolved oxygen concentrations, and the barge directed by radio to the required position. The introduction of continuous automatic monitoring stations measuring oxygen levels would assist in the operation.

The oxygenation equipment would be 'skid-mounted' so that it could be removed from the barge and transferred by motor vehicle to another site if it were required in an emergency, for example, to alleviate the effects of an accidental spillage of pollutants having high oxygen demands.

Pollution Budgets and Consent Standards

The condition of the Tideway has been restored using, on a somewhat piecemeal basis, first the concepts of the Pippard Report, and later the recommendations of the Third Report of the Royal Commission on Environmental Pollution. A plan for future management is now necessary, so that the quality of the estuary can be maintained, and present and future dischargers are equitably treated regarding the standard permitted to them for their effluents (Wood *et al* 1979b).

The concept of the pollution budget (p93) identifies it as the maximum pollution load that can be accepted if the required standards for the estuary are to be achieved. For the Thames the budget is probably best defined in terms of oxygen deficiency, but other parameters (such as temperature) will probably require consideration in future.

Before an acceptable pollution budget can be established, two factors must be considered: (*a*) the potential ecological quality of the section of the estuary;

and (*b*) the preparedness and ability of the community to finance the necessary measures for environmental protection and improvement. If the potential ecological quality is not high, or the public not able (or prepared) to pay for the improvements necessary to achieve the full ecological potential, then the estuarine quality objective will be lower, and a larger pollution budget available for the wastes discharged.

When the budget has been established, it must be shared amongst all dischargers in an equitable manner, the share of each being used to determine the quality required from their discharges. The Royal Commission offered two simple criteria for estuarine water quality as a suitable basis for establishing the budget: the estuary should be able to support the passage of migratory fish at all states of the tide, and should also be able to support on the mud bottom the fauna required for the maintenance of sea fisheries.

Ecological Considerations

As with most large estuaries, the Thames ranges from the salt water of the sea, through brackish water, to the fresh water above the tidal limit at Teddington Weir. The estuary can therefore be divided into three zones based on salinity (figure 92).

Section 1: Teddington to London Bridge. Although the effect of the tides is observed throughout the Thames estuary up to Teddington (at London Bridge there is an average rise and fall of 7 m between high and low water), under normal conditions the river between Teddington and London Bridge

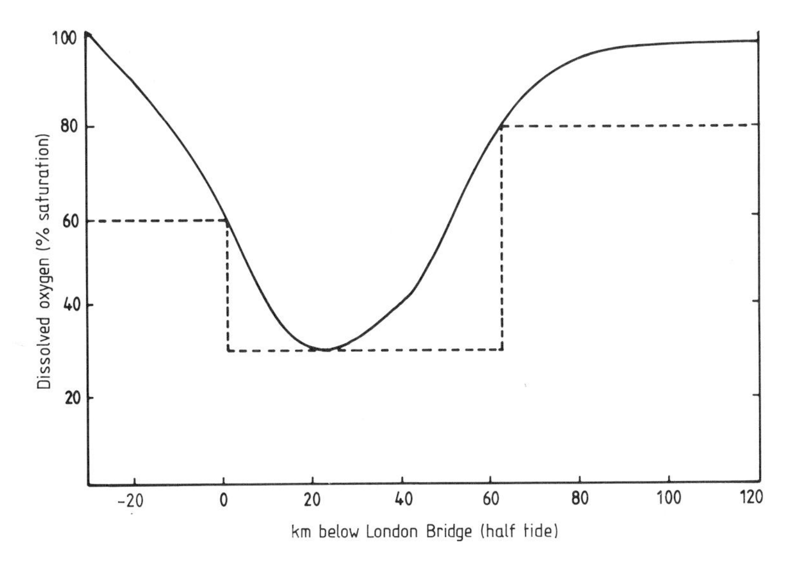

Figure 92. Dissolved oxygen pollution budget for the Thames Tideway.

can be regarded as non-saline. Freshwater criteria have therefore been applied to the zone, and it has been decided that this stretch of the Thames can be regarded as a coarse fishery, and has therefore been graded as class 2 (see table 33).

Section 2: London Bridge to Canvey Island. In a seaward direction downstream of London Bridge, the influence of sea water incursion becomes increasingly significant. Flora and fauna become typical of mesohaline zones of estuaries, and in this region the criteria of the Royal Commission are applied to set the quality objectives. As has already been stated, this requires a minimum of 30% saturation of dissolved oxygen as a quarterly average. Statistically, however, it can be shown that this is compatible with a 95 percentile limit of 10% saturation with dissolved oxygen.

Section 3: Canvey Island to the Sea. Most marine organisms are tolerant to salinities as low as 18 parts per thousand (10 000 mg/l chlorion), and this condition is satisfied a few kilometres downstream of Gravesend. However, the physical conditions are too harsh there for these organisms owing to the restricted width of the river which emphasises the strong currents. Suitable environmental conditions for the whole life-cycle of marine organisms are first found just upstream of Canvey Island, so that this has been chosen as the first point downstream at which this classification begins.

Quality Objectives

Using the above classifications, the quality objectives for the different zones of the Tideway are as shown in table 23. The limits shown in the table can be used to determine pollution budgets for different parameters such as deficiency of dissolved oxygen, temperature and toxic substances inhibitory to fisheries.

Establishment of Pollution Budgets

The pollution budget is the maximum pollution load that the estuary can accept whilst still maintaining the selected environmental conditions of quality. If the dissolved oxygen concentration limits of table 23 are plotted, a curve such as the broken line in figure 92 is obtained. In practice the estuary could never conform to such sharp changes in dissolved oxygen levels as occur between the river sections, so that a curve such as the continuous line of figure 92 is more likely to be achieved. Since this curve conforms to quality requirements, the oxygen deficiency represented by the area above the curve is acceptable, and therefore becomes the pollution budget in respect of oxygen. The total oxygen depletion from all discharges must not be such that they cause a greater deficiency than this represents.

Having established the pollution budget, it must be shared equitably amongst all dischargers. First, an allowance must be made for non-point

Table 23. Objectives for the different zones of the Tideway.

Stretch of estuary	Objective
Teddington to London Bridge	(i) The minimum percentage air saturation with dissolved oxygen should be at least 40% in the case of 95% or more of the samples analysed. (ii) The water should be non-toxic to fish.
London Bridge to Canvey Island	(i) The minimum percentage air saturation with dissolved oxygen should be at least 10% in the case of 95% or more of the samples analysed. (ii) The quarterly minimum average percentage saturation with dissolved oxygen should not fall below 30%.
Canvey Island to seaward limits	(i) The quality should be suitable for the whole life cycle of marine organisms, including fish. (ii) The minimum percentage air saturation with dissolved oxygen should be at least 60% in the case of 95% or more of the samples analysed.

source discharges such as surface run-off. In the Thames area much of the sewerage is on the combined system, and an allowance of 10% of the budget will probably suffice. A decision must then be made regarding what proportion of the remainder is to be used for established discharges, and what allowed for future effluents. If insufficient is allowed, then, when new discharges have used all the reserve, a re-proportioning exercise will be necessary for a new distribution, and a revision of standards will be required. On the other hand, if too large a fraction is held in reserve, it will be defeating the whole object of control by pollution budgets.

In order to share the budget equitably, it is recognised that eventually the public will pay for the treatment of all discharges. They pay for sewage treatment works and also, in the cost of manufactured goods, that fraction of the price attributable to effluent treatment costs. It is assumed, therefore, that all dischargers are entitled to a fraction of the budget, proportional to the pollution load which they would discharge if no treatment were given. Detailed consideration must be given to each case; for example, the only pollution load acceptable for the calculation of the share of the budget will be the minimum which is possible, by good design and housekeeping at industrial premises making discharges.

The measure for assessing the pollution load is the effective oxygen load (EOL), which is defined as

$$\mathrm{EOL} = F \times 10^{-6}(1.45B + 4.57N),$$

or, more approximately,

$$\text{EOL} = F \times 10^{-6} \times \tfrac{3}{2}[B + 3N],$$

where $B = 5$ day BOD (ATU) (mg/l), $N =$ oxidisable nitrogen (mg/l), and $F =$ flow rate (m^3/day). The share of the budget is given by:

$$\frac{\text{EOL of any discharge}}{\Sigma\text{EOL of all discharges}} = \frac{\text{Area over curve allotted as budget share}}{\text{Total area over curve}}.$$

Having allocated part of the area above the curve as a share of the budget (figure 92), then by mathematical modelling or similar techniques, it will be possible to determine to what standard the effluent must be treated so that it does not deplete the dissolved oxygen by more than the area allocated. In the case of tributaries discharging to the estuary, the potential pollution load at the confluence must for this purpose be regarded as the sum of the loads of sewage works (and other effluents) discharged to the tributary. Although it will be necessary to allow a portion of the budget for storm sewage discharges, because of the sporadic nature of their occurrence they cannot be allocated a share in the same way as other discharges.

The budget for discharges for the Tideway (after an allowance of 10% for non-point discharges) is shared between the different discharges, as shown in table 24. The present consent conditions for Thames Water Authority sewage works discharging to the middle reaches of the tidal Thames are shown in table 25. By meeting their consents these works use up the proportions of the total pollution budget shown in table 26; their allocated shares are also given. It is evident that some works will use more and some less than their allowable proportions, but taken together they just use their allocated share.

Table 24. Distribution of pollution budget.

Discharge to estuary	Allowable proportion of pollution budget (%)	Actual amount of pollution budget used at present (%)
Freshwater rivers	40	26
Thames Water Authority sewage works	53	53
Other sewage works	4	6
Industrial discharges	3	15

All figures after 10% allowance for non-point discharges.

Table 25. Consent conditions to be applied to sewage treatment works in the middle reaches of the tidal Thames.

Discharges	Flow (m^3/day)	BOD (mg/1)		Ammonia nitrogen	
		Mean	95 percentile	Mean	95 percentile
Beckton	1 000 000	10	22	2.0	6
Crossness	568 000	12	25	7.5	16
Riverside	95 000	20	31	30	44
Long Reach	200 000	20	28	40	53

Table 26. Distribution of pollution budget (Thames Water sewage treatment works).

Discharges	Allowable proportion of pollution budget (%)	Actual amount of pollution budget used at present (%)
Mogden	9.2	9.7
Kew	0.6	0.8
Beckton	22.4	7.3
Crossness	12.9	12.2
Riverside	3.1	5.7
Long Reach	4.8	17.7
Total	53.0	53.4

Dissolved oxygen is the main factor which controls environmental conditions in the estuary, and so its budgeting is of prime importance. Pollution budgets for other parameters can, however, be determined in a similar manner.

Flood Defence

The Thames originally had a natural flood plain to receive the extra volume of water when the river level rose. This land has since been developed and protected by flood defences along the banks of the estuary, but now there is an increasing risk that these defences may be overtopped by exceptionally high tidal conditions. Over the past 162 years high water levels have risen by 1.22 m at London Bridge, equivalent to an increase of 0.76 m per century (figure 93). This change is due to several causes (Horner 1978). The Institute of Oceanographical Sciences has shown it to be due to a rise in mean sea level of 0.36 m per century, and also to increases in tidal amplitude affected by such factors as the dredging and embankment of the estuary.

This rise in sea level can be accommodated and does not pose a difficult problem; it is the 'storm surge' which presents a serious flooding risk, and could cause a disaster, especially by the extensive disruption of transport and other services in London (figure 94). This would involve expenditure of hundreds of millions of pounds, and remedial work would take a very long time. The surges are caused by deep meteorological depressions which approach the west coast of Ireland and can move in a north-easterly direction with a velocity of about 100 km/h. The difference in pressure of 30 mbar resulting from the depression could raise sea level by 0.3 m over a circle 1600 km in diameter. If this 'hump' of water moves from the Atlantic to the shallower waters of the continental shelf, sea level could be raised by 0.6 m. In the event of this mass of water moving down from Scotland in a south-easterly direction and into the North Sea, particularly in coincidence with a northerly gale, the resulting rise in sea level could be 2.5 m or more in the Thames estuary. If this 'storm surge' is maintained at high water, then the increased height would be superimposed on the normal high tide and a dangerously high water level will be the result.

Various schemes have therefore been devised to protect London from such a disaster by a tidal barrier, and the final choice lay in the construction of a defence barrier at Silvertown, 14 km below London Bridge. Tidal structures can be of three kinds:

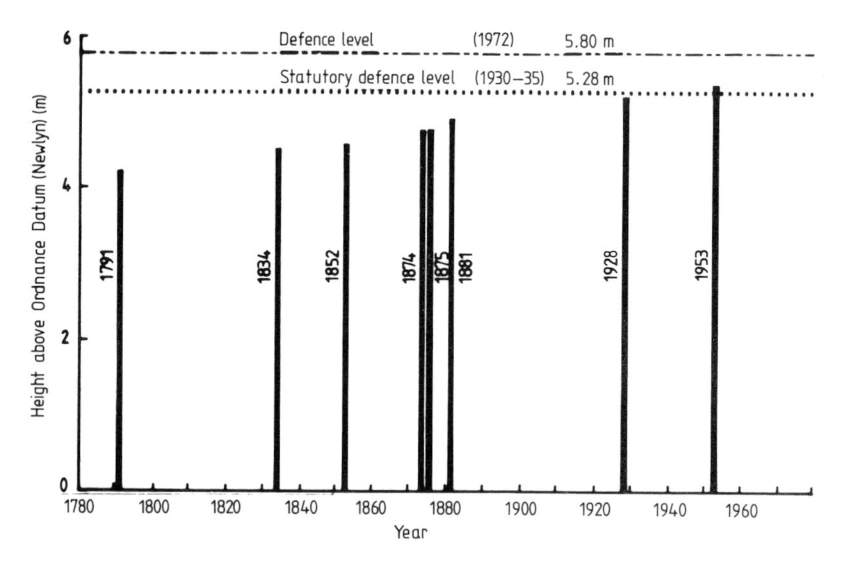

Figure 93. Increasing high tide levels at London Bridge.

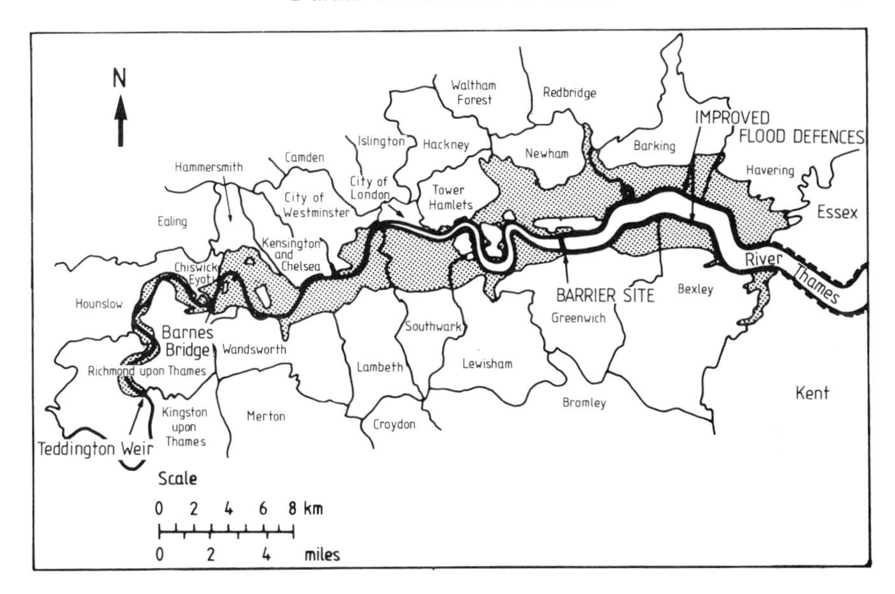

Figure 94. The maximum area likely to be affected by a surge tide flood.

Figure 95. A tidal barrier to prevent flooding. © Greater London Council.

(i) A tidal barrier, which is put into operation only during periods when there is a risk of flooding, and so offers the minimum hindrance to navigation and is likely to cause only an insignificant and temporary effect in respect of pollution and siltation.

(ii) A barrage, which is a fixed structure which will permanently hinder navigation, and can have significant pollution and siltation effects.

(iii) A half-tide barrier is a compromise, being a structure which impounds the river flow only during low tide, and which is likely to have only minimal effects on pollution and siltation.

The installation at Silvertown will be a tidal barrier of the rising sector gate type shown in figure 95. To allow the passage of ships of up to 20 000 tonnes gross registered tonnage, four openings 61 m wide, and six smaller ones of 31.5 m, each with sector gates, are being constructed (figure 95). Flood defences downstream (and to some extent upstream) of the barrier have been raised and strengthened.

Environmental Effects of the Thames Barrier

A bill was promoted by the Greater London Council in 1972 to obtain powers to build and operate the barrier, and this received Royal assent in August of that year. It is to be noted that these powers are designated for the operation of the structure as a tidal barrier, the environmental effects of which are likely to be insignificant. To operate it as a half-tide control system (figure 96) would require another Act of Parliament. There are, however, advantages in controlling the water levels upstream at half-tide levels. There would be an operational advantage in the control of surge tides, and the increased water surface area above the barrier would offer opportunities for sports and other amenities at low water. To some, the absence of the uncovered mud banks at low tide would be deemed an advantage. Although the barrier has locks to allow passage of shipping when the gates are raised, it would undoubtedly be restrictive to navigation if it were used for half-tide control.

A tidal barrier on the Thames for the purpose of flood defence is likely to cause only insignificant effects on the environmental quality of the river simply because it will be used so infrequently.

Studies have been carried out by the Water Pollution Research Laboratory into the effects of a half-tide barrier on the chemical condition of the river, using a time-dependent model (Greater London Council 1969, 1970). The conclusions reached were that the half-tide operations would produce a slight deterioration in river quality generally in the region of the barrier. At the time it was not expected that the present river quality standards would be achieved, so the results are probably not applicable to today's conditions. These studies did not, of course, take account of the effects of storm sewage discharges which could affect the condition of the river upstream of the barrier very severely. However, the use of oxygen injection, as already described, would rectify these conditions.

Undoubtedly the barrier, when used to maintain half-tide control, could cause considerable increases in water temperature although these would quickly disappear when it was opened.

Studies using a large-scale physical model at the Hydraulics Research Station indicate that there would be a tendency for some suspended matter to move upstream towards the barrier and cause siltation to occur just downstream of it. The exact effects of the operation of the barrier for half-tide control on the Tideway in its present condition could be obtained by mathematical modelling, but this hardly seems to be worthwhile because when the barrier is built, assuming the necessary statutory sanctions are obtained for half-tide operation, the effects could be obtained by direct monitoring.

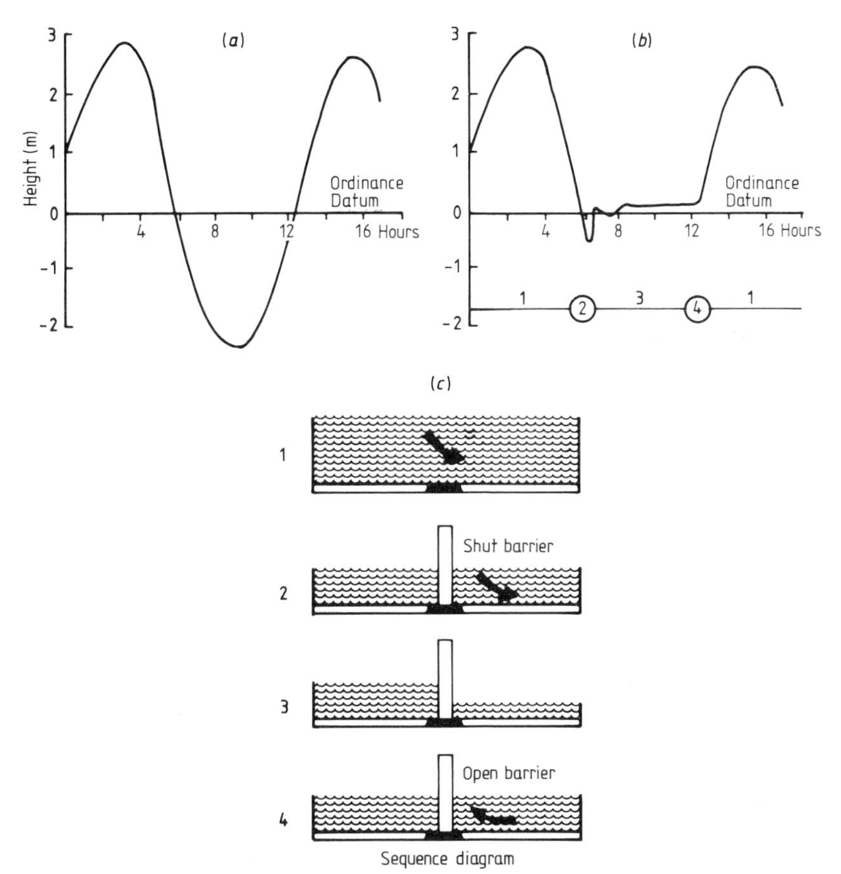

Figure 96. A half-tide control system. (*a*) Open river tide profile; (*b*) tide control profile upstream of the barrier; (*c*) tide control sequence diagram.

The ranges of salinity at different points along the river with and without half-tide control are given in table 27. Changes in salinity can be very important to fish. Very few freshwater species can survive in salinities greater than 5 g/1000 g, and few marine species in salinities below 18 g/1000 g. The barrier, however, is sited at the oligohaline/mesohaline boundary where salinities are unsuitable for fish of both the above kinds, and only euryhaline species thrive. However, as the changes in salinity throughout a tidal cycle are much greater than those produced by half-tide operation, the latter are not likely to have significant effects.

Table 27. Change in range of salinity by half-tide barrier control.

Position below London Bridge (km)	Range of salinities (g/1000 g NaCl)	
	Without control	With control
58	20.5–26.1	20.3–26.3
45	13.2–20.3	14.8–20.3
31	7.5–13.5	7.8–13.5
23	4.5–10.0	4.8–9.4
20	3.3–8.7	3.9–7.7
14	1.1–6.1	1.5–5.1
5.7	0.3–2.8	0.5–1.5
0.3	0.2–1.4	0.4–0.5

The main concern about half-tide control would undoubtedly be the interruption of fish migration, both upstream and downstream, by reason of the localised thermal barrier occurring just upstream of the closed barrier. Wildfowl could also be affected, simply because some of the mud flats of bays would not be exposed to provide sources of food such as tubificid worms. These latter could also be affected by changes in the size of particulate matter in the substrate, as a result of the effect of the barrier on the distribution of suspended matter and siltation.

Looking to the Future

The quality of the Thames Tideway now conforms to all that can be expected of it in the light of its present uses. Indeed, bearing in mind the financial restraints, it may even be a struggle to maintain the present quality to the end of the century. Undoubtedly, any future changes in quality requirements will be related to the increased demand for supplies of potable water which are likely with the projected population growth and improved standards of living.

The river water is too saline in most parts of the estuary for abstraction for potable supply, but the potential of the Upper Thames for this purpose has recently been seen to be related indirectly to the quality of the Tideway. The Thames Conservancy Act of 1932 required that abstraction from the freshwater Thames must be controlled so that a quantity of at least 0.77×10^6 m^3/day remains to flow into the Tideway. If, however, the level of water in the reservoirs of the Thames Valley falls at greater than a specified rate during a drought, then the upland flow at Teddington can be reduced according to a 'chart', but it must never fall below 0.23×10^6 m^3/day. These provisions were necessary in the first three-quarters of the century, when the flow remained at the 'statutory minimum' for a few weeks and the tidal river became anaerobic.

In 1976, however, the United Kingdom faced the worst drought for 200 years; reservoirs were depleted, and in many cases empty. A dispensation was therefore sought to use a larger amount of the water of the Upper Thames for abstraction, and for several weeks the whole river was used. Even the leakage through Teddington Weir was pumped back, and the flow over Molesey Weir (the first weir downstream of the abstraction points) was reversed, so that all the fresh water could be abstracted. The effect on the Tideway was merely that it became narrower, and the salinity pattern changed due to the greater inland penetration of sea water. As has been pointed out on p124, the water quality remained at the required level, although some species of fauna moved further upriver for the period. It proved, however, that at least for a short period of time the Tideway could maintain a satisfactory condition without any added upland flow. The quantity of water flowing into the Tideway, even under the statutory minimum conditions of 0.77×10^6 m^3/day, is approximately 48% of the amount required for use in Greater London for domestic and industrial abstraction.

The experience gained during the 1976 drought made it clear that with forward planning a less restrictive form of 'chart' could allow greater abstraction in the early days of a drought, so that the extreme measure of total abstraction would seldom be required. In addition, use might be made of the freshwater tributaries such as Beverley Brook as sources of potable supply. The disadvantage of the use of such streams is that they are composed largely of high-quality sewage effluent, the retention time of which in the river can be counted in days, instead of the minimum of a month usually required before such water can be abstracted for reservoir storage. This deficiency could no doubt be overcome by increasing the storage time in reservoirs and, if necessary, by the use of more sophisticated treatment such as activated carbon to remove those substances and organisms usually removed by river retention.

It may be that in the future some attention will be given to the re-use of the large flows of effluent from Beckton and Crossness. Both are too saline at present (probably due to in-leakage of river water into sewers) for consideration as raw material for potable supply, but this could be remedied if the need became great enough. These effluents might be used for recharging underground aquifers, subject to certain reservations, in two possible ways. The drawing down of the water table under Greater London has resulted in saline penetration from the estuary, with the result that some boreholes adjacent to the Tideway cannot now be used. Recharge with high-quality effluents could act as a barrier between the estuarine water and that of the aquifer. A further possibility would be to use Beckton's effluent to recharge the chalk north of the river. It is the policy in this area to consider recharge only with water treated to a potable quality standard; it would therefore be necessary to introduce some form of advanced treatment before the recharge was undertaken.

In considering the re-use of water, far greater care is necessary returning it to underground sources than to surface waters. If a stream becomes polluted, often the situation is remedied by allowing the slug of pollution to be carried downstream, carefully monitoring it to ensure that the water is not used incorrectly, when eventually the pollutant will become diluted and the stream will recover. However, if an aquifer becomes polluted, for example by a spillage of chemicals or oil on exposed chalk, removal is extremely difficult, if not impossible. The only remedy is to sink boreholes and pump out the water, which creates a 'cone of depression' into which the pollutant is drawn and pumped to waste.

If consideration is given to aquifer recharge, therefore, one must be absolutely certain of the harmlessness of the water used. It may be possible to do this by a procedure currently adopted to ensure that the water abstracted from the Upper Thames and Lee can be safely ingested (Wood and Richardson 1979). If the water is to be ingested over a lifetime, special care must be taken to ensure that it is satisfactory. Although EEC and WHO standards are observed for the supply of potable water, where such water is derived from rivers containing sewage effluents, special precautions may be necessary, since sophisticated artefacts may have been introduced by way of industrial discharges of unknown toxicity. It is impossible to set precise standards to limit such compounds, unless their compositions are known. Therefore, since there are no standards for any known compound which set a limit of less than 0.1 μg/l, any suspected organic compounds present in excess of this amount are regarded as potentially harmful until they are identified and cleared. It is not possible to determine these chemicals quantitatively by direct analysis, such as by mass spectrometry, until their nature has been ascertained, but even then some are so obscure that methods of analysis must be developed before they can be quantified. The mass balance of such materials may be derived by quantifying the amounts from all sources, and then determining what will be their total quantity at abstraction points. If it exceeds 0.1 μg/l, the effects of lifetime ingestion of this dose will be studied, and if they are unacceptable, the source will be eliminated.

No immediate need is envisaged for the use of effluents for aquifer recharge, but should it ever be contemplated, a study on the above lines would be necessary. Probably, however, by the time such need arises, the mother necessity will have produced many inventions for improved water treatment.

Appendix I

River Pollution and Purification

A river may receive pollution from many sources (figure 97). Flows such as sewage and industrial effluents which enter at a single point are called 'point discharges' and are therefore subject to control; those which enter from a wide and undefined area, such as run-off from fields, are known as 'non-point discharges', and their control is more difficult. Polluting materials fall into two broad classes: those which can be broken down by natural purification processes and are therefore biodegradable, and those which are non-biodegradable.

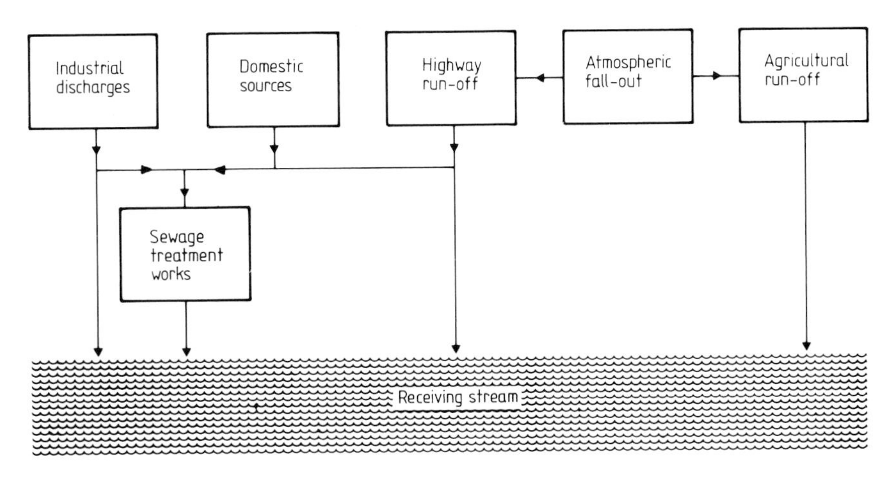

Figure 97. Sources of pollution of a watercourse.

Non-biodegradable Pollutants

Non-biodegradable pollutants are almost certain to be artefacts which do not form part of the natural biological cycle of rivers. They include heavy metals or other toxic inorganic substances, and organic chemicals which may be toxic to the biota, or have chemical structures not readily biodegradable, or both. In the latter class, pesticides and polychlorinated biphenyl compounds probably cause the most serious problems. Such substances must be controlled by ensuring that they are either completely prevented from entering water courses, or else only allowed in regulated quantities so that at no point in the river do they reach concentrations which will affect the biota adversely. Examples of the toxicity of some non-biodegradable substances are given in table 28.

Table 28. Examples of the toxicity of some non-biodegradable substances.

Substance	Concentration lethal to fish (mg/l)
Aldrin	0.02
BHC (gammexane)	0.035
DDT	0.1
2:4D	500
Sodium pentachlorophenate	0.04
Parathion	0.2
Toxaphene	0.005
Phenols	1–20
Cyanide	0.04–0.1
Cadmium chloride	0.017
Copper	0.03
Mercury	0.01

Biodegradable Pollutants

Most biodegradable materials are organic in nature and they themselves, or their breakdown products, form substrates for the biota of the stream. Bacteria are the main organisms concerned, and they can be broadly classified into *autotrophs,* which require only simple substances for their metabolism, using carbon dioxide or its compounds to synthesise carbohydrates and protein; and *heterotrophs,* which cannot use carbon dioxide in this way, and must obtain carbon from organic compounds. Autotrophs can be *chemosynthetic*, deriving energy from oxidation of inorganic substances such as ammonia (*Nitrosomonas* spp), or *photosynthetic,* when they derive energy from light. Those bacteria which require free oxygen for growth are designated aerobic. Others (anaerobic bacteria) cannot survive in presence of air and derive oxygen from that present in such salts as sulphates and phosphates.

There is an intermediate stage between aerobicity and anaerobicity. Aerobic decomposition of biodegradable organic material will proceed by using the dissolved oxygen in a stream until its level falls to about 5% saturation.

Thereafter, nitrate and nitrite are used as sources of oxygen (the process is called denitrification), preserving 'anoxic' conditions until all nitrate and nitrite is used, after which anaerobic conditions develop and sulphate is converted to sulphide (p82). When denitrification processes occur, the product is elementary nitrogen which escapes to the atmosphere as an innocuous gas; when anaerobic conditions prevail, sulphide is formed, which is extremely lethal to the river biota, and can give rise to hydrogen sulphide in the riverside atmosphere.

Aerobic Biodegradation

Although the uptake of oxygen by carbonaceous pollutants in a stream follows the biochemical kinetics described by Monod (1941), it can be represented by a physico-chemical pseudo-first-order reaction, provided the oxygen concentration is above the very low level at which it becomes rate-limiting. If L_0 is the total oxygen demand of a single substance, then the rate of oxidation is

$$-\mathrm{d}L/\mathrm{d}t = kL_0, \tag{1}$$

where k is the rate constant. On integration, this gives

$$y = L_0\,(1 - \mathrm{e}^{-kt}), \tag{2}$$

where y is the oxygen uptake after time t, and L_0 is the oxygen demand at the outset. It would be most unusual to have only one biodegradable carbonaceous substance present; the numbers are usually very large and each has its own rate constant:

$$y = L_0\,[1 - (p_1\mathrm{e}^{-k_1t} + p_2\mathrm{e}^{-k_2t} \ldots + p_n\mathrm{e}^{-k_nt})], \tag{3}$$

where p_n is the fraction of the load attributed to the nth constituent.

By a process of curve fitting, it has been found for the Tideway that oxygen uptake by carbon-containing compounds could be best represented by

$$y = E_\mathrm{c}[1-\{(1-p)\mathrm{e}^{-0.23t} + p\mathrm{e}^{-0.046t}\}], \tag{4}$$

where E_c is the effective oxygen demand (or the total oxygen demand of the polluting demand in respect of carbon), and from this equation the demand can be calculated after time t.

The constituents are assumed to be capable of being divided into 'fast' carbon with a k value of 0.23 day $^{-1}$, and 'slow' carbon with a k value of one fifth of that rate, in the proportion of $(1-p)$ to p. E_c can be related to the five-day BOD in which *Nitrosomonas* are suppressed by the use of allyl thiourea (Wood and Morris 1966):

$$E_\mathrm{c} = AB, \tag{5}$$

where B is the BOD and A is a constant related to p, as shown in the table below.

	p	A
Settled sewage	0	1.45
Non-nitrified secondary effluent	0.5	2.22
Nitrified secondary effluent	0.75	3.03

The temperature variation of k is given by (Barrett *et al* 1978):

$$k_T = 0.183\,(1.05)^{(T-15)}. \tag{6}$$

In the oxidation of carbonaceous wastes a multitude of intermediate products will be formed, but the final products will be carbon dioxide and water. The oxidation of nitrogen-containing wastes involves three stages:

(*a*) the decomposition of organic nitrogen compounds to ammonia;
(*b*) the oxidation of ammonia to nitrite by *Nitrosomonas* bacteria; and
(*c*) the oxidation of nitrite to nitrate by *Nitrobacter* bacteria.

The stoichiometric equation for (*b*) is:

$$NH_4^+ + 1\tfrac{1}{2}O_2 = 2H^+ + H_2O + NO_2',$$

whereby it is apparent that one part of ammonia nitrogen requires 3.43 parts of oxygen for oxidation to nitrite. The corresponding equation for oxidation of nitrite to nitrate is

$$NO_2' + \tfrac{1}{2}O_2 = NO_3'.$$

Combining these equations, the oxidation of ammonia to nitrate is

$$NH_4^+ + 2O_2 = NO_3' + 2H^+ + H_2O,$$

whereby one part of ammonia nitrogen requires 4.57 parts of oxygen for oxidation to nitrate†. These proportions are in fact slightly lower in both cases because ammonia is used by some organisms to synthesise their protein. Again Monod has devised kinetics based upon the growth rates of the bacteria involved, but it is found in practice that a pseudo-first-order reaction gives predictions which agree with observed data. The oxygen uptake can be represented by

$$x = 4.57N\,[1 - e^{-Kt}], \tag{7}$$

where

$$K = 0.39 \text{ at } 20\,^{\circ}\mathrm{C}$$

† One part of *ammonia* requires 3.76 parts of oxygen (p112).

and

$$N = N_{amm} + [AB/U_C \times N_{org}],$$

where N_{amm} is the ammonia originally present, $[AB/U_C \times N_{org}]$ is the ammonia generated from organic nitrogen compounds, and AB/U_C is the fraction of the total substrate decomposed in the time t. The corresponding effective oxygen demand with respect to nitrogen, E_N, is given by $4.57N$.

The growth rates of the nitrifying organisms (*Nitrosomonas* and *Nitrobacter*) are very much influenced by temperature and pH. The generation time varies from about 30 hours in summer conditions, to about two weeks in winter (compared with about 20 minutes for the carbon-oxidising bacteria), and they are considerably inhibited by lowered pH values (Downing 1968), as occasioned by the build-up of carbon dioxide in the gas phase in contact with the liquid.

The temperature gradient of K is

$$K_T = 0.26\,(1.09)^{(T-15)} \tag{8}$$

where T = temperature (°C).

Effective Oxygen Loads

E_C and E_N are expressions of concentrations of pollutants in discharges, but a much more useful measure is that of the total quantity or load of oxygen that an effluent will consume. This is expressed as the effective oxygen load (EOL), and is obtained as:

$$\text{EOL} = F \times 10^{-6}(E_C + E_N) \text{ tonnes/day},$$

where F = flow (m^3/day) and E_C and E_N are the effective oxygen demand (mg/l). For settled sewage this becomes

$$\text{EOL} = F \times 10^{-6}\,[1.45B + 4.57N]$$

which is sometimes simplified to

$$\text{EOL} = F \times 10^{-6} \times 1.5\,[B + 3N],$$

where B is the BOD (ATU) and N is the total oxidisable nitrogen concentration.

Denitrification

Denitrification is a process whereby nitrate in a stream can be used as a source of oxygen for oxidation of organic materials under conditions of low dissolved oxygen concentration. Even in rivers (and water storage reservoirs), where the level of dissolved oxygen is high in the bulk of the water, denitrification occurs to a significant extent, probably by reaction with the organic material in the benthic deposits, where dissolved oxygen is lowest. As has already been mentioned, the presence of nitrate can be beneficial in conditions such as those of the Tideway, but in other rivers it can be regarded as a pollutant. Where the water is used as a source for potable supply, nitrates in concentrations above 100 mg/l have been suspected of causing methaemoglobinaemia in bottle-fed infants, and of being potentially carcinogenic by reacting to form *N*-nitrosamines in the human gut. Nitrates can also contribute to eutrophication (over-enrichment of nutrients) which can be a problem in some watercourses due to excessive growth of algae and enteromorpha. The denitrification processes described above can therefore be beneficial in some rivers. The growth kinetics of the organisms responsible for denitrification are greatly dependent on temperature.

Oxygenation of Watercourses

Oxygen can enter a stream by three different means:

(1) By advection, i.e. addition from tributaries and other sources such as lakes and the sea.

(2) By photosynthesis, i.e. the conversion of carbon dioxide and water to sugars and oxygen in presence of light in plants containing chlorophyll:

$$6CO_2 + 6H_2O = C_6H_{12}O_6 + 6O_2.$$

(3) By solution from the atmosphere at the surface.

The rate of entry of oxygen into water is given by:

$$dC/dt = K(C_s - C),$$

where dC/dt = rate of uptake of oxygen, and K = constant. The quantity C is the concentration of dissolved oxygen in the water at the time of aeration, and C_s is the saturation value, that is, the maximum concentration of oxygen that can be achieved, at a specific temperature, concentration of dissolved substances, and the given partial pressure of oxygen in the gas being used as the source of oxygen. The value of C_s decreases with increased temperature (p73) and the concentration of salts, and increases according to Henry's law with the partial pressure or concentration of oxygen in the gas in contact with the liquid.

Hence if 'pure' oxygen is used instead of air (which contains only about 20% oxygen), C_s is about five times as great. Hence when oxygen is introduced into a stream where the concentration is C, the rate of solution, $K(C_s - C)$, will be much greater when pure oxygen is used instead of air as the source of oxygen. Pure oxygen injection into streams containing high levels of pollution, such as from storm sewage (p143), can sometimes be the only method of supplying sufficient oxygen to meet the demand. The constant K is usually called the constant of proportionality, or overall absorption coefficient. It increases with the surface area A, and is inversely proportional to the volume of water, V, undergoing aeration.

It is more convenient, however, to use a measure of aeration rate which is independent of the surface area and volume of the liquid undergoing aeration; this can be obtained as follows:

$$K \propto A/V$$

hence

$$K = f(A/V)$$

(f is a constant, and $f = K\,(V/A)$. The dimensions of K are (t^{-1}), of V/A (l), and hence of f (lt^{-1}). The parameter f gives us a measure of rate of aeration independent of the surface area and the volume of liquid undergoing aeration. This is called the exchange coefficient, and is measured in centimetres per hour (cm/h).

The Effects of Synthetic Detergents

Synthetic detergents, like all surface-active substances, have the property of concentrating at the surface of a liquid, often as a monomolecular layer, according to the Gibbs equation:

$$\Gamma = \frac{\mathrm{d}\gamma}{\mathrm{d}c}\frac{C}{RT}$$

where Γ = surface excess (mass per unit area of surface); γ = surface tension; C = concentration of surface-active material in the bulk of the liquid; R = gas constant, and T = absolute temperature.

For a given concentration of surface-active material (C), the surface excess Γ will be proportional to $\mathrm{d}\gamma/\mathrm{d}c$, that is, to the rate of lowering of surface tension with increased concentrations of the surface-active substance. This differential coefficient can be large for synthetic detergents so that they spread readily on the surface layer, and can be regarded as partially sealing the surface against the entry of oxygen from the atmosphere to the bulk of the liquid.

Oxygen Balance in Streams

The level of dissolved oxygen in a watercourse will be the result of equilibrium between that present and what is being introduced, and that which is being removed by the river's own self-purification processes in oxidising biodegradable pollutants. One of the most important aspects of pollution control must therefore be to control the levels of polluting material in discharges so that after their oxygen requirements have been met, there is sufficient left in solution to support the biota and allow the water to be used for the purposes required of it, as shown in table 33.

Appendix II

Principles of Sewerage, and Sewage Treatment and Disposal

Sewerage Systems

Sewerage systems are of two general kinds. Separate systems involve the collection of domestic wastes and industrial discharges into 'foul sewers' which convey the 'foul sewage' for treatment at a sewage works. In this system, surface drainage (runoff from roofs and roads) is conveyed by separate sewers for discharge (generally without treatment) to a watercourse, care having been taken to install oil interceptors at appropriate points. Combined systems of sewerage receive both foul sewerage and surface drainage and convey it to a sewage works. New drainage systems seem to favour separate systems which are not so much affected by storms.

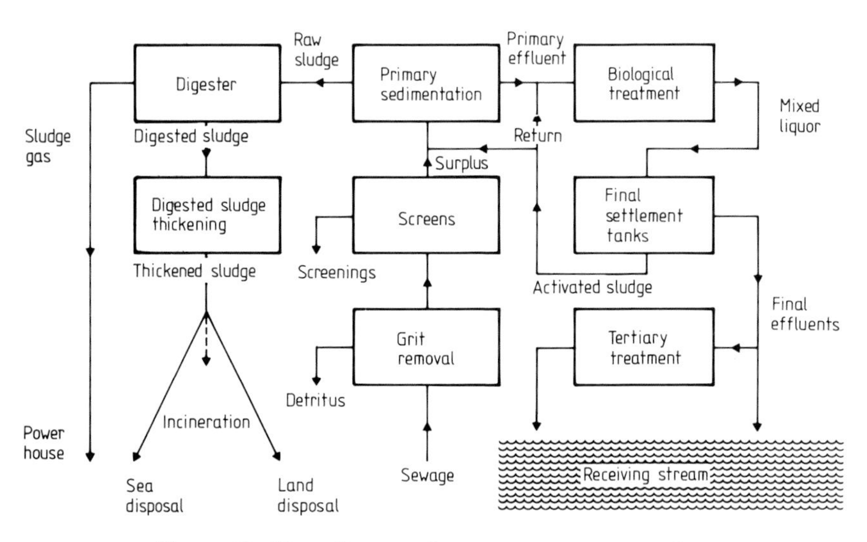

Figure 98. Flow diagram of a sewage treatment works.

Since the public's use of water varies throughout the day, so does the flow of sewage arriving at a treatment works. In dry weather this usually follows a constant daily pattern referred to as the diurnal variation of flow.

The processes involved in sewage treatment are extremely numerous and those selected for a particular works will depend upon many circumstances; it is the purpose here merely to describe processes common to the larger sewage works in Greater London, to which reference has already been made (figure 98). The sewage arriving at the works (table 29) is composed of both domestic sewage and trade discharges, but except in the case of Riverside works, where there is separate treatment of trade effluents, the exercise of a competent trade effluent control allows the mixed sewage to be treated with few problems. Sewage treatment can be divided into various stages, each leading to a further degree of refinement of effluent quality (table 30).

Table 29. Percentage of industrial effluent in London's sewage.

Sewage works	Total flow (10^6 m^3/day)	Percentage of industrial effluent
Beckton	1.00	6.0
Beddington†	0.087	3.7
Crossness	0.59	6.6
Deephams†	0.20	9.0
Hogsmill/Surbiton†	0.064	1.5
Kew	0.040	4.8
Long Reach	0.20	6.2
Mogden	0.51	8.0
Riverside	0.098	20.3
Worcester Park†	0.016	3.1
Sutton†	0.009	—

† Discharges to tributaries of the Tideway.

Table 30. Stages of sewage treatment (see figure 98 for flow diagram).

Preliminary treatment	Removal of grit, rags and large objects
Primary treatment	Separation of solid matter by sedimentation
Secondary treatment	Oxidation and removal of dissolved, colloidal and fine suspended matter, usually by biological treatment
Tertiary treatment	Further treatment to extend secondary treatment

Preliminary Treatment

Sewage contains detritus (grit) washed in from roads. In sewers this is kept in suspension by maintaining the flow rate above 1 m/s, the minimum velocity required for self-cleansing of the sewers. The detritus must be removed at an early stage of sewage treatment, as otherwise it will cause excessive abrasion of pumps and machinery. The sewage is therefore passed through detritus tanks, where the flow velocity is reduced to 0.3 m/s, which allows the detritus to settle. The tanks are usually of the constant-velocity type which, because of their cross sectional shape, maintain the required linear velocity for a range of flow rates.

Other debris which could affect machinery includes wood, rags and other relatively large objects. Most commonly these are removed by screens; primary screens remove the largest objects, and these are followed by secondary screens which remove smaller objects. Screenings can be burnt, or taken for disposal to a solid waste tip. Alternatively, they can be treated by macerators, which shred the screenings and return them upstream of the screens, so that large remaining objects pass a second time, and eventually become small enough to pass without causing difficulties to the next stage of treatment. Sometimes comminutors are installed, which shred the screenings *in situ*.

Primary Sedimentation

Sewage is passed slowly and continuously through primary sedimentation tanks to remove as much solid matter as possible by sedimentation. The solid matter is called 'raw sludge' and is passed to the sludge digestion plant. The supernatant liquid, called 'primary effluent' or 'settled sewage' is then given biological treatment.

Primary sedimentation tanks are designed to provide about two hours' retention of sewage at the maximum flow rate (three times dry weather flow) with a surface loading (sewage flow per unit surface area of tank) of 1.25 m/h at the maximum flow rate. The efficiency of solids removal is reduced when the sewage contains 'surplus' activated sludge, because of the low density of these solids. Typical removal efficiencies are 60–75% in respect of suspended solids and 45–60% for BOD; the sludge usually contains about 3% solids if surplus activated sludge is included. At large treatment works the tanks tend to be rectangular with a bridge mechanism which moves along the tank, lowers the scraper blade to the floor at the outlet end, and sweeps the settled sludge back to a sump at the inlet end, from which it is pumped out. A scum board is usually moved back through the surface at the same time to remove oil and grease.

Biological Treatment

Primary effluent is next treated to oxidise and remove organic matter in suspension and solution. This process is, in effect, basically the same as the process of bacterial oxidation in rivers; the difference is that because ultimately the oxidation rate depends upon the number of organisms present, biological treatment involves carrying out the process with a very high concentration of organisms, and thereby reduces the time for removal of pollutants. There is obviously an advantage in removing as much of the oxygen-consuming load as possible; the remainder will be oxidised in the receiving stream, causing a commensurate removal of dissolved oxygen.

Biological treatment therefore involves bringing the primary effluent into contact with a high concentration of bacteria, and at the same time supplying sufficient oxygen for the process to be effected. There are many different methods of carrying out the process, but the main two are by the use of biological filters, or the activated sludge process.

Biological Filters

This was the earliest form of biological treatment first used by Dibdin in his 'one-acre coke bed'. The name 'filter' is inappropriate, since the plant consists of a bed packed with material with a large surface area, such as coke (plastics filters are now also used), which allows the primary effluent to come into contact with these surfaces. When the filter has matured, a film covers the surface and contains a large population of the required bacteria. Sewage is applied to the bed through spray jets (usually the hydrostatic head of sewage turns the rotor arms supporting the jets), and in passing through the interstices, reacts with the biological film and with oxygen dissolved from the air. Filters are circular or rectangular, and can produce high-quality effluents, from which a substantial part of the organic material (BOD) and ammonia have been oxidised. They are less susceptible to shock loads than activated sludge, but tend to be less able to oxidise ammonia in colder weather. The effluent passes into humus tanks to remove solid matter before it is discharged to the receiving stream. Filters are subject to seasonal sloughing of the film.

Activated Sludge Process

At a large sewage treatment works the area required for biological filters would be too great, and the activated sludge process is more common. In this process a sludge is developed which supports a high bacterial concentration; this is comparable with the film of the biological filter. Primary effluent and activated sludge are mixed ('mixed liquor'), and passed through tanks which are aerated either by diffusers on the bottom which produce small air bubbles, or by mechanical aerators which rotate on the surface, and introduce oxygen by creating large areas of interface. At the end of the process the purified liquid (final effluent) is separated in final tanks from the activated sludge, which is recirculated. The rate of removal of BOD and ammonia depends upon the concentration of bacteria, and hence of the activated sludge solids, and the time of retention in the aeration tank. The concentration of sludge solids that can be accommodated is limited by the quantity of oxygen which can be made available to them. 'Pure' oxygen can be used to enable the process to operate with a very high suspended solids concentration (and hence short retention times). Unfortunately, this causes problems when it is necessary to achieve a high degree of removal of ammonia because of the build-up of carbon dioxide in the aerating gas, which lowers the pH and inhibits the growth rate of *Nitrosomonas* (Wood *et al* 1976). Generally, using air as a source of oxygen, and a plant capable of giving about eight hours' retention, with a 'mixed liquor' suspended solids concentration of about 3000 mg/l, an effluent with BOD of 10 mg/l or less, and ammonia nitrogen of 1 mg/l can be achieved, and this is satisfactory for discharge to many streams. The process builds up activated sludge so that a fraction must be wasted; this is usually returned and added to the sewage upstream of the primary sedimentation plant.

Tertiary Treatment

In some circumstances where, for example, an effluent is to be discharged into a very high-quality river (such as class 1B; see table 33) and where there is little dilution available in the river, a better effluent may be required than is possible by secondary treatment. The effluent is then 'polished' by a form of further treatment. This can take the form of removal of solids by micro-strainers, by sand filters which remove solids and extend the oxidation process, or by passing over land which achieves the same result. It is sometimes necessary to remove nitrate; this can be done by reacting the nitrate in the final effluent with a carbon source such as methanol, in the presence of denitrifying bacteria (Barnard 1973). A sounder method, however, is to introduce an 'anoxic zone' by keeping mixed liquor in suspension without aeration in the first part of the aeration tanks of the activated sludge system, so that the nitrate present in the internatant liquor of the returned activated sludge is used to oxidise part of the BOD load of the primary effluent (Cooper *et al* 1977). Further treatment to a very high standard can be incorporated (such as by activated carbon, if necessary) to remove organic constituents in solution.

Effluent Quality Standards

Reference has been made to the need for a standard for effluents commensurate with achievement of the quality objectives of the rivers into which they discharge. This can be done for each determinand of the river quality objective by mathematical modelling, or by a simple calculation:

If Q_0 = the flow rate upstream of the discharge, C_0 = the quality of water upstream of the discharge; Q_1 = the flow rate of effluent; C_1 the quality required of the effluent C_2 = the quality of water required downstream of the discharge;

$$Q_0C_0 + Q_1C_1 = (Q_0 + Q_1)C_2,$$

and

$$C_1 = \frac{(Q_0 + Q_1)C_2 - Q_0C_0,}{Q_1}$$

i.e.

$$\begin{matrix} Q_0C_0 \searrow & \\ & (Q_0 + Q_1)C_2 \rightarrow \\ Q_1C_1 \nearrow & \end{matrix}$$

Sludge Digestion and Disposal

The raw sludge leaving the primary sedimentation plant is very unpleasant and is difficult to thicken (i.e. to increase the solids concentration). It is usually digested, i.e. fermented anaerobically at about 33 °C, when not only is the resulting digested sludge less obnoxious and easier to thicken, but at large works sufficient sludge gas (66% methane, 34% carbon dioxide) is produced to supply the energy requirements of the works (sometimes with a little to spare). The sludge is retained for about 21 days in the digester when about 45% of organic matter is removed, and the total solid matter reduced by 33%.

Digested sludge can be disposed of in a number of ways. A very small number of works burn it in incinerators, but this is not usually a preferred method, and can be costly. The commonest method, especially for inland works, is to dewater the sludge in lagoons or drying beds to about 12½% solids or more, and then to spread it on land as a soil conditioner and low-grade fertiliser. Some problems are experienced due to the heavy metal content (especially of cadmium), but the Ministry of Agriculture, Fisheries and Food (MAFF) have laid down guidelines for an acceptable rate of spreading.

In London about 60% of digested sludge is disposed of by shipping to the Barrow Deep in the outer estuary. Specially constructed ships† are used which, when the tanks are filled with sludge, maintain a hydrostatic head of sludge level above the sea water. As the sludge runs out and the ship rises in the water, the head is still maintained, so that the discharge can be made entirely by gravity. The disposal operation is required to be completed within a prescribed area marked by four buoys. Such disposal is only permitted under a licence from MAFF, and is controlled under the Dumping at Sea Act 1974, and the Oslo Convention of 1972. Such a method of sludge disposal

† Usually called 'sludge vessels'.

has been carried on in the area of the Barrow Deep since about 1890, and after the instructions of Dibdin in 1893, the estuary has been regularly monitored as far out as the spoil ground. There has never been any evidence of ecological damage as the area is rapidly swept seawards by strong currents (figure 99).

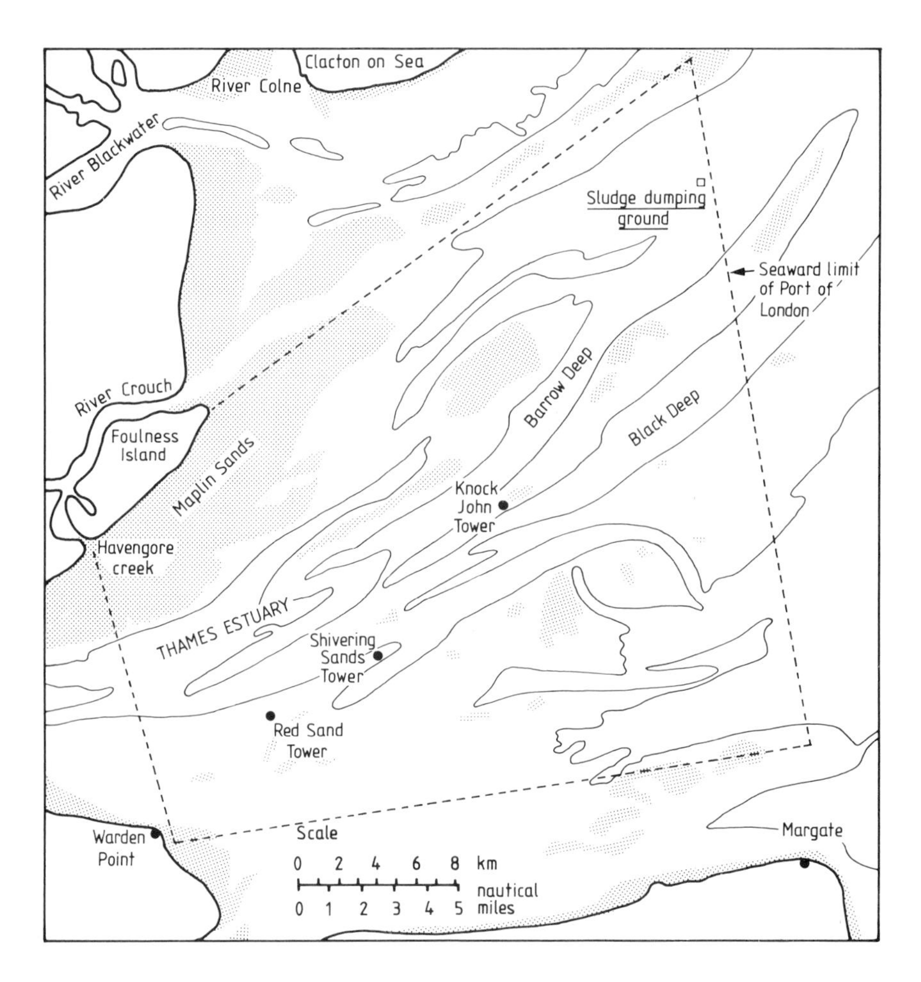

Figure 99. The Thames estuary, showing the position of the Barrow Deep, site of sewage sludge disposal since 1890.

Appendix III

Legislative Procedures for Pollution Control

Legislation involving the control of pollution is of two types in the United Kingdom: by statute or under common law. The latter is the legislation available to all members of the public, and can be invoked to recover damages (such as for the loss of fish stocks) or to seek injunctions for the restraint of damaging activities. Regulatory bodies such as the water authorities are usually concerned with statutory legislation.

Statutory legislation is to all intents and purposes the same in England and Wales, but in Scotland it is considerably different. That concerned with pollution control falls into two categories: (i) to control discharges to sewers, and (ii) to protect water resources.

Control of Discharges: Trade Effluents

Each householder in the UK has the right, under the Public Health Act 1936, section 34, to have his private sewer connected to the main sewerage system, provided that no noxious matter is caused to flow into it. In the case of industrial discharges (usually called 'trade effluents') these may only be discharged to the public sewer subject to certain restrictions. The Public Health (Drainage of Trade Premises) Act 1937 required that after an 'appointed day' trade effluent could only be discharged after a consent was obtained from the appropriate regulatory authority (now the water authority) which could require that undesirable constitutents should be eliminated or reduced to a satisfactory level, and similar control could be exercised in respect of temperature and acidity. Discharges made before the 'appointed day' were exempt. The Public Health Act 1961 brought the pre-1937 discharges under control by 'direction', in that they were required to conform to certain standards of temperature and pH. In addition, some discharges were made by 'agreements' with the then regulatory authority, and if there was no termination clause, the agreement could be rescinded only if the volume of discharge was changed. The Public Health Act 1961 also empowered the authority to charge for the receipt and treatment of trade effluents.

In London, however, the legislation was different. Originally the discharge of trade effluents was dealt with by the London County Council (General Powers) Act 1953, but by the passing of the London Government Act 1963, the national enactments applied broadly, although the 'appointed days' were different. By the recent enactment of the Control of Pollution Act 1974, of which the section 43 relating to trade effluents is now in force, practically all the anomalies of previous enactments were rationalised. All trade effluents now require a consent for discharge, this being generally required to stand for two years before revision. The new enactment allows the water authority to revise the consent in a shorter period, but if it does so, the discharger may be reimbursed in compensation claims.

The water authority has two purposes in sampling trade effluents: in the one case for protecting the purification system, and in the other for charging.

Control samples are taken to ensure:

(i) that the discharge does not present a hazard to men working in sewers or at sewage works;

(ii) that the discharge will not harm the fabric of the sewer; and

(iii) that the discharge will not interfere with the biological and other processes at sewage works.

It would be very unwise to have common national standards of limitations for substances discharged in trade effluents, since conditions vary so widely, and each case must be examined on its own merits. Examples of the kinds of restraint applied in parts of the London sewerage system are given in table 31.

Rating samples of trade effluent are taken so that, together with the measurement of the volume of trade effluent, the total pollution load will be known and a charge may be levied. The charge represents the best attempt that can be made to assess the cost of reception and treatment. The concentrations of those constitutents requiring treatment at a sewage works are determined, compared with the corresponding average concentration in the authority's domestic sewage, and the charge related thereby to the average cost of sewage treatment. The National Water Council's charging formula is given in table 32.

Protection of Water Resources

There has been legislation for control of river pollution since the Rivers Pollution Prevention Act of 1876. The first modern act was the River Boards Act 1948, which established river boards to exercise this control. These were later consolidated as river authorities under the Water Resources Act 1963, and their functions were absorbed by regional water authorities under the Water Act 1973. The regulatory authorities have powers granted under the

Table 31. Consent conditions for discharge of trade effluents to sewers in London.

Restricted substance	Maximum permitted (mg/l) (typical)
Settleable solids	250–1000
Cyanide as CN	5–20
Sulphide as S	1–10
Ammonia as NH_3	50–100
Sulphate as SO_3	1000–1500
Available chlorine as Cl	10–50
Available sulphur dioxide as SO_2	1–10
Formaldehyde as HCHO	5–1000
Grease and/or oil	0–500
Tarry matter	0–500
Cadmium as Cd	1 or 0.1 kg/day
Chromium as Cr	15
Copper as Cu	5
Lead as Pb	10
Arsenic as As	10
Nickel as Ni	5
Silver as Ag	5
Zinc as Zn	5
pH	6–11
Temperature	43 °C
Organic solvents	Prohibited
Petroleum spirit	Prohibited
Calcium carbide	Prohibited

Table 32. Charging for trade effluents.

Charging formula $R + V + (O_t/O_s)\,B + (S_t/S_s)/S$ per m^3 trade effluent, where

R = One third of average cost per m^3 of receiving sewage into the sewers and conveying it to the sewage treatment works

V = average cost of treatment by primary sedimentation of 1 m^3 of sewage

O_t = chemical oxygen demand (mg/l) of trade effluent

O_s = chemical oxygen demand (mg/l) of average sewage

B = average cost per m^3 of biological treatment of sewage

S_t = total suspended solids (mg/l) settleable in half an hour at pH 7.0 in the trade effluent

S_s = total suspended solids (mg/l) settleable under similar conditions in average sewage

S = average cost per m^3 of sewage received for sludge treatment and disposal

Table 33. Suggested classification of river quality.

River class	Quality criteria	Remarks	Current potential uses
	Class-limiting criteria (95 percentile)		
1A	(i) Dissolved oxygen saturation greater than 80%. (ii) Biochemical oxygen demand not greater than 3mg/l. (iii) Ammonia not greater than 0.4mg/l. (iv) Where the water is abstracted for drinking water, it complies with requirements for A2† water. (v) Non-toxic to fish in EIFAC terms (or best estimates if EIFAC figures not available).	(i) Average BOD probably not greater than 1.5mg/l. (ii) Visible evidence of pollution should be absent.	(i) Water of high quality suitable for potable supply abstractions and for all other abstractions. (ii) Game or other high-class fisheries. (iii) High amenity value.
1B	(i) Dissolved oxygen greater than 60% saturation. (ii) BOD not greater than 5 mg/l. (iii) Ammonia not greater than 0.9 mg/l. (iv) Where water is abstracted for drinking water, it complies with the requirements for A2† water. (v) Non-toxic to fish in EIFAC terms (or best estimates if EIFAC figures not available).	(i) Average BOD probably not greater than 2 mg/l. (ii) Average ammonia probably not greater than 0.5 mg/l. (iii) Visible evidence of pollution should be absent. (iv) Waters of high quality which cannot be placed in Class 1A because of high proportion of high-quality effluent present or because of the effect of physical factors such as canalisation, low gradient or eutrophication. (v) Class 1A and Class 1B together are essentially the Class 1 of the River Pollution Survey.	Water of less high quality than Class 1A but usable for substantially the same purposes.
2	(i) Dissolved oxygen greater than 40%	(i) Average BOD probably not greater than	(i) Waters suitable for potable supply after

	(ii) BOD not greater than 9 mg/l. (iii) Where water is abstracted for drinking water, it complies with the requirements for A3† water. (iv) Non-toxic to fish in EIFAC terms (or best estimates if EIFAC figures not available).	(ii) Similar to Class 2 of RPS. (iii) Water now showing physical signs of pollution other than humic coloration and a little foaming below weirs.	(ii) Supporting reasonably good coarse fisheries. (iii) Moderate amenity value.
3	(i) Dissolved oxygen greater than 10% saturation. (ii) Not likely to be anaerobic. (iii) BOD not greater than 17 mg/l‡.	Similar to Class 3 of RPS.	Waters which are polluted to an extent that fish are absent or only sporadically present. May be used for low-grade industrial abstraction purposes. Considerable potential for further use if cleaned up.
4	Waters which are inferior to Class 3 in terms of dissolved oxygen and likely to be anaerobic at times.	Similar to Class 4 of RPS.	Waters which are grossly polluted and are likely to cause nuisance.
X	Dissolved oxygen greater than 10% saturation.		Insignificant watercourses and ditches not usable, where objective is simply to prevent nuisance developing.

Notes

(a) Under extreme weather conditions (e.g. flood, drought, freeze-up), or when dominated by plant growth, or by aquatic plant decay, rivers usually in Classes 1, 2 and 3 may have BODs and dissolved oxygen levels, or ammonia content outside the stated levels for those Classes. When this occurs the cause should be stated along with analytical results.

(b) The BOD determinations refer to 5 day carbonaceous BOD (ATU). Ammonia figures are expressed as NH_4.

(c) In most instances the chemical classification given above will be suitable. However, the basis of the classification is restricted to a finite number of chemical determinands and there may be a few cases where the presence of a chemical substance other than those used in the classification markedly reduces the quality of the water. In such cases, the quality classification of the water should be downgraded on the basis of the biota actually present, and the reasons stated.

(d) EIFAC (European Inland Fisheries Advisory Commission) limits should be expressed as 95% percentile limits.

† EEC category A2 and A3 requirements are those specified in the EEC Council Directive of 16 June 1975 concerning the Quality of Surface Water intended for Abstraction of Drinking Water in the Member States.

‡ This may not apply if there is a high degree of re-aeration.

Rivers Prevention of Pollution Acts of 1951 and 1961, to control pollution of streams and watercourses. Under these acts it is an offence knowingly to permit poisonous, noxious or polluting matter to enter a watercourse. Exceptions are made in certain cases, the most common being of effluents discharged to streams with the consent of the regulatory authority. The consent of a water authority must be obtained:

(*a*) for the making of a new or altered outlet for discharge of trade or sewage effluent to a stream; or

(*b*) for the making of a new discharge of effluent.

The water authority will issue a consent to discharge based mainly (in the case of sewage and trade effluents) on BOD, suspended solids, temperature and sometimes ammonia. Consents can be reviewed at two-year intervals, or sooner, by the agreement of both parties, and in the case of a dispute, the matter is referred to the Secretary of State.

The Water Resources Act 1963 gave river authorities powers in respect of the control of abstractions of water and the issuing of licences therefor. It also became an offence to discharge polluting, poisonous or noxious matter into underground strata by means of a 'well, borehole or pipe'. Unfortunately, this did not cover discharges by percolation through the soil, such as from a broken oil pipe. The only recourse in this situation was under the Water Act 1945 if a water supply was at risk.

The Control of Pollution Act 1974, which is now partly in force, will strengthen all previous legislation, bringing control over discharges to practically all water resources. It will also remove a limitation of the previous legislation, whereby only the water authority or the Attorney General can institute legal proceedings under the acts; when the appropriate sections of the new legislation come into force, any individual has the right. The new Act will also require water authorities to keep registers of analyses, etc, taken in the course of their pollution prevention activities, and these must be made available for inspection by the public at reasonable times. When the water authorities were formed they 'inherited' consents drawn up by different authorities, sometimes with different approaches, especially in respect of the degree of compliance of an effluent with its consent. An effluent which complies *on average* with the Royal Commission standard of BOD 20 mg/l, suspended solids 30 mg/l will only comply with BOD 30 mg/l, suspended solids 45 mg/l for 95% *of all samples* taken.

In order to present a uniform system of consents, the National Water Council first drew up a scheme (National Water Council 1979) for the classification of rivers according to their uses, and attached the chemical and other parameters to enable the appropriate quality to be achieved (table 33). It is thereby possible to determine by a simple mathematical model, the effluent standards required to meet these river quality objectives on a 95 percentile basis (p164). Since it is not sensible to put a stream into the class required of it, if it is known that the stream cannot meet the standards because effluents of an unsatisfactory quality will be discharged into it, and no finances are available to improve these effluents, then a special designation is given. For example, a stream would be classified as 3/U2/U1B, meaning that at present the best quality that can be achieved (with good management at the works discharging unsatisfactory effluents) is 3, but that when finances are available the effluents must be upgraded to achieve river quality first of 2 and then later of 1B.

The Control of Pollution Act 1974 (Part I) provides waste disposal authorities (usually county councils) with powers to control the deposition of solid wastes. However, in order that the drainage (leachate) from solid waste tips does not pollute water resources, the waste disposal authority must seek the agreement of the appropriate water authority, before granting the site licence required by anyone wishing to make a deposition of solid waste.

Legislation

This list gives all Acts of Parliament mentioned in this book governing the River Thames and the control of pollution, and the provision of water and sewerage services. These Acts are arranged in chronological order with page references to the text.

Legislation to ensure that the quality of the upper Tideway was not allowed to deteriorate . . ., early in C20, 64
Thames Conservancy Act, 1932, 149, 178
Public Health Act, 1936, 95, 167
Public Health (Drainage of Trade Premises) Act, 1937, 95, 167
Water Act, 1945, 172
River Boards Act, 1948, 87, 168
Rivers (Prevention of Pollution) Act, 1951, 87, 172
London County Council (General Powers) Act, 1953, 168
Public Health Act, 1961, 95, 167, 168
Rivers (Prevention of Pollution) Act, 1961, 87, 172
London Government Act, 1963, 168
Water Resources Act, 1963, 168, 172
Port of London Act, 1968, 92
Bill to build and operate the Thames Barrier, 1972, 146
Water Act, 1973, 168
Control of Pollution Act, 1974, 94, 95, 168, 172, 173
Dumping at Sea Act, 1974, 165

Appendix IV
Mathematical Modelling

A mathematical model is a tool which enables a water quality planner to simulate the behaviour of a river when conditions are changed, by varying in an appropriate way the inputs to what is in effect a complicated algebraic equation. Models can be of two types:

(i) *stochastic models*, which use values of a determinand obtained historically and calculate, by interpolation or extrapolation, the changes in the determinand when the value of another determinand changes, and

(ii) *deterministic models*, which attempt to simulate actual changes by modelling each process involved by an equation or function.

Models may also be *steady state*, when they are used to predict a situation resulting from a change of conditions persisting over a reasonably long period of time (of say a month); or *time-dependent*, when the immediate effect of changes lasting only a short period of time can be evaluated. Models used for policy planning of water resources tend to be deterministic, steady state models.

The model for the Thames Tideway (Gameson *et al* 1955) considers the estuary as a barrier-free tidal river divided into a number of segments, each being regarded as homogeneous. It takes account of mixing coefficients of the constituents in each section. The basic equation is:

$$\frac{\mathrm{d}}{\mathrm{d}x} EA \frac{\mathrm{d}C}{\mathrm{d}x} - V \frac{\mathrm{d}C}{\mathrm{d}x} - KC = 0,$$

where E = mixing coefficient; A = cross sectional area; V = freshwater velocity seawards; K = rate constant for the material under examination; x = distance along the estuary; and C = concentration of constituent. The model is based on pseudo-first-order kinetics of the kind described in Appendix I by 'fast' and 'slow' BOD, organic nitrogen and by ammonia. The effect of nitrate as a source of oxygen is recognised. Basically, the model is the classical one for diffusion combined with advection. Despite the fact that an average exchange coefficient (5.1 cm/h) for ambient temperature is assumed for the whole estuary, and that no account is taken of the oxygen depletion of resuspension of benthic material at high spring tides (Wood 1970), the model gives good agreement between predicted and observed levels of dissolved oxygen and, except during the third quarter of the year when the water temperature is above average, the prediction is good for nitrate and ammonia concentrations.

It is, of course, necessary to write the model into a programme for a computer, the calculations being far too involved for manual resolution. The computerised model can then be used as in the examples given in the text. An input can be made to the model at a location on the estuary in terms of load of BOD, organic nitrogen and ammonia, and the depreciation in river quality which would result if the discharge were in fact made, can be determined. Similarly, the effect of discharges of conservative substances (those which pass through the estuary unchanged) such as chlorides, can be shown in respect of their distribution throughout the estuary. Since the mathematical model will show the effects of different discharges, it enables the planner to determine where to make the input to have the best cost-effectiveness.

Appendix V

Hydrography of the River Thames

The Upper Thames

The Upper Thames is a river classified on the National Water Council system as 1B (see table 33) for much of its length, and is therefore suitable for potable supply abstraction, game and other high-class fisheries, and is of high amenity potential. It is precluded from being in class 1A because of the large quantities of effluents discharged into it, even though they are of high quality, and even the lengths of the river in class 2 are, of course, capable of supporting good coarse fisheries. Although the river receives a considerable amount of sewage effluent (figure 100), its quality has always been strictly controlled and maintained in good condition (Barclay 1963), so that it is used for the abstraction of large quantities of water to be treated for potable supplies.

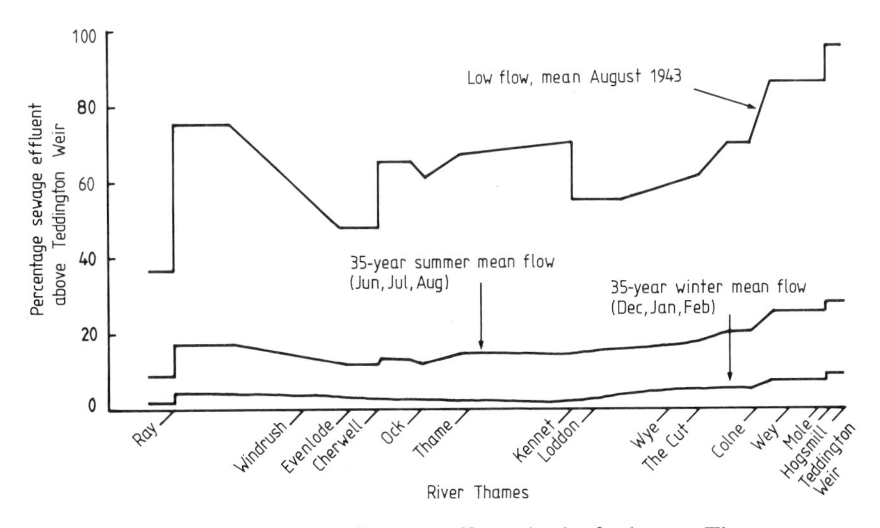

Figure 100. Percentage of sewage effluent in the freshwater Thames.

Over the first three-quarters of this century the flow of fresh water to the Tideway from the Upper Thames has undoubtedly been essential to provide a measure of dissolved oxygen, and to dilute and transport impurities seaward. With the growth of Greater London in this period, demands for potable water increased so that it became necessary to restrict abstraction from the upper river to ensure that sufficient flow passed to the Tideway. The Thames Conservancy Act of 1932 required that after abstraction at least 0.77×10^6 m^3/day of water must be left to flow to the Tideway.

The Thames Tideway

The tidal River Thames varies in width and depth in a seaward direction (figure 101). It is subject to a tidal cycle, and the heights of tides (averages of springs and neaps) are shown in figure 102. As the tide ebbs and flows, a particular point on the water surface will move forwards and backwards over the range shown in figure 103(*b*). This movement (tidal excursion) is over a length of approximately 15 km in 12.5 hours, which means that if a particular geographical location was sampled, it would have the water of that length of river flowing past it in half a day, and the quality might be subject to considerable change. It is customary, therefore, to adopt a 'half-tide correction'. If the time of sampling relative to high water is known, then it is possible, using a curve of the nature of figure 103(*a*), to determine the position in the river occupied by the sample at the time of half-tide, and to assume that the quality of the sample was the quality of the river at that

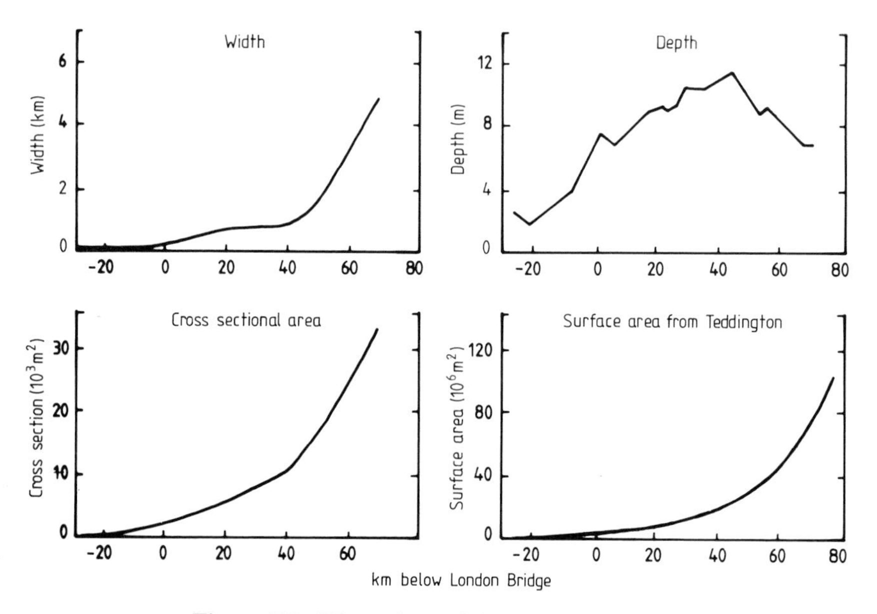

Figure 101. Dimensions of the Thames estuary.

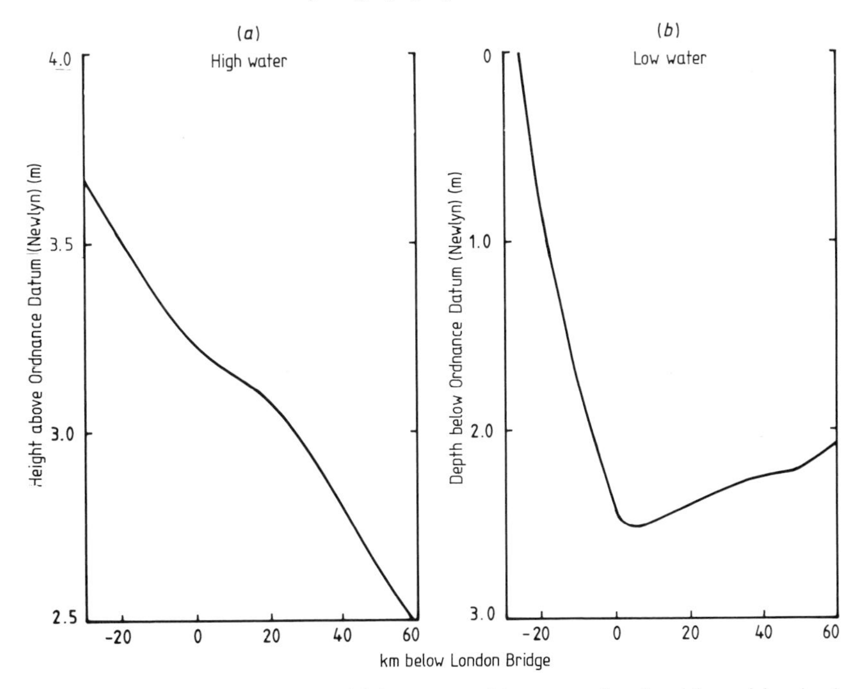

Figure 102. Variation in average high water and low water levels with position in the Thames estuary (mean of spring and neap tides).

point. Such a system allows a determination of water quality to be made at all points in the river at half-tide. Such quality data can be compared for different sampling runs, regardless of the points in the tidal cycle at which the samples are taken.

Not only does a given point of the river water oscillate with the tide, but it is also subject to the seaward movement caused by the upland flow of fresh water. This rate of movement will be proportional to the rate of upland flow. The average rates of freshwater flow are given in table 34. The third (summer) quarter is usually the period which has the lowest average flow, and it is conventional to compare river conditions for this quarter. When predictions are made by mathematical modelling (see Appendix IV), third quarter conditions with the statutory minimum flow of 0.77×10^6 m^3/day are used. The rate of seaward movement for different rates of freshwater flow can be seen in figure 104.

The salinity of the Tideway (measured in grams of sodium chloride in 1000 grams of water—g/1000 g) varies from almost zero (0.05) at Teddington, to 34 where it meets the North Sea (figure 105). Since sea water is more dense than fresh water, where the two meet there is sometimes a tendency for the latter to flow over the former to form a salt wedge, as can also be found in the estuary of the River Tees. In the Thames Tideway, however, the salinity

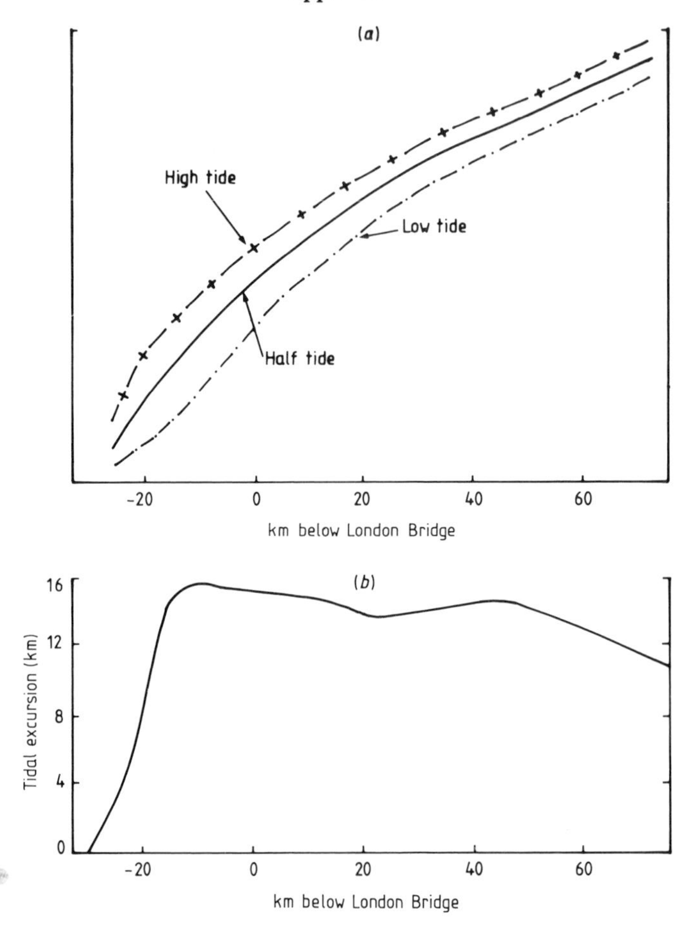

Figure 103. (*a*) Corresponding position in the estuary at low, high and half tides. (*b*) Tidal excursion at different positions in the Thames estuary.

Table 34. Average rates of upland flow of freshwater over Teddington Weir, 1883–1962.

Quarter	Flow (10^6 m^3/day)
First (January–March)	10.36
Second (April–June)	4.69
Third (July–September)	1.96
Fourth (October–December)	6.32
Average for year	5.83

is practically constant over the whole cross sectional area of the basin, due to its configuration. The river passes through a series of serpentine convolutions in which the water is driven to change sides at each bend; this induces mixing and leads to uniform salinity. The whole pattern of salinity is moved

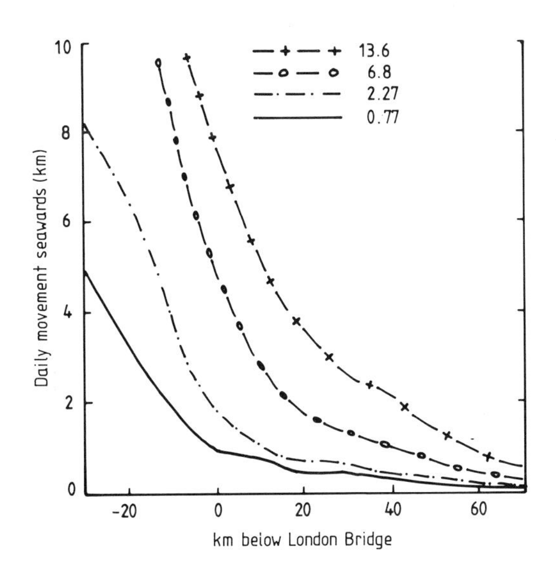

Figure 104. Daily movements for different upland flows (10^6 m^3/day).

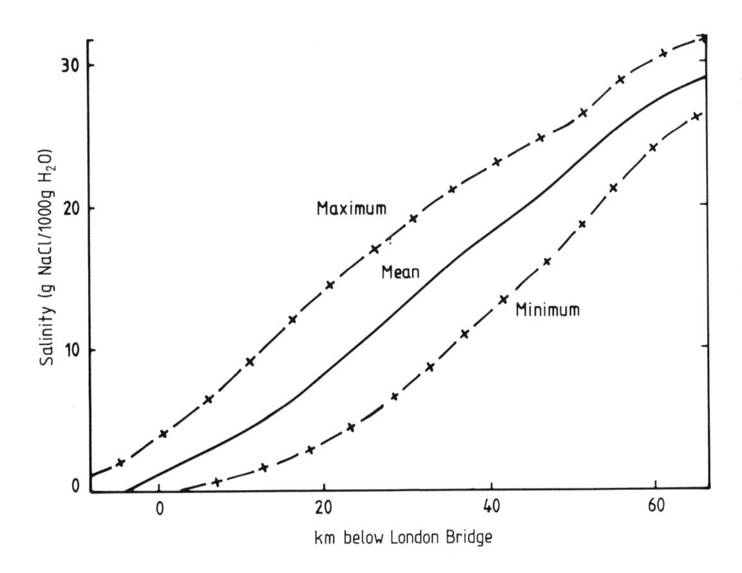

Figure 105. Variation in salinity of the Thames estuary.

up and down the river by variations in the rate of upland flow. To give some idea of the movement, it is conventional to indicate the points where the water is slightly brackish (a concentration of chlorion, Cl^-, of 1000 mg/l), and that regarded as the limit of sea water penetration (100 mg/l chlorion).

Freshwater Sources Entering the Thames Estuary

The principal tributaries of the tidal Thames are shown in table 35. The largest is the River Medway. Water is abstracted from the freshwater Upper Thames and from the River Lee upstream of its confluence with the Tideway, the water being used as the raw material for treatment for potable supply. The average quantities abstracted daily from these rivers are 1.34×10^6 m^3/day from the Upper Thames, and 0.27×10^6 m^3/day from the River Lee.

Table 35. Tributaries of the Tidal Thames.

River	Confluence with Thames	
	N or S bank	Distance from London Bridge (km)
Upper Thames	—	30.2 above
Crane	N	24.3 "
Duke of Northumberlands	N	23.8 "
Brent	N	21.8 "
Beverley Brook	S	12.8 "
Wandle	S	10.6 "
Ravensbourne	S	7.2 below
Lee	N	11.0 "
Roding	N	18.7 "
Beam	N	22.7 "
Ingrebourne	N	24.2 "
Darent and Cray	S	29.1 "
Mardyke	N	29.4 "
Ebbsfleet	S	39.7 "
Medway	S	70.4 "

Apart from the freshwater input from the tributaries, the Thames estuary receives a considerable flow of sewage effluents (table 36). The flow from the largest of these (Beckton) was once greater than that of the River Medway and was therefore the largest tributary of the Thames. The flows from the two sources are now similar.

Table 36. Major sewage works discharging to the tidal Thames.

Sewage works	Water authority	North or south bank	Position below London Bridge (km)	Flow (10^6 m^3/day)	Consented load EOL (tonnes/day)
Mogden	Thames	N	−24.2	0.51	39
Kew	"	S	−19.6	0.04	2.3
Worcester Park†	"	S	−12.8†	0.016	1.2
Sutton†	"	S	−12.8†	0.009	0.9
Beddington†	"	S	−10.4†	0.087	12
Deephams†	"	N	+11.3†	0.20	17
Beckton	"	N	+18.3	1.00	69
Crossness	"	S	+22.6	0.59	71
Riverside	"	N	+24.1	0.098	33
Long Reach	"	S	+31.2	0.02	61
Marsh Farm	Anglian	N	+43.3	0.023	11
Swanscombe	Southern	S	+36.7	0.001	0.3
Northfleet	"	S	+39.9	0.006	1.1
Gravesend	"	S	+45.0	0.008	5
Stanford	Anglian	N	+51.5†	0.005	0.8
Basildon	"	N	+57.9†	0.019	3
Pitsea	"	N	+57.9†	0.006	1
Canvey	"	N	+59.8	0.005	2
Benfleet	"	N	+62.8†	0.007	1.2
Southend	"	N	+66.3	0.043	26

† Effluents discharged to tributaries of the Tideway.

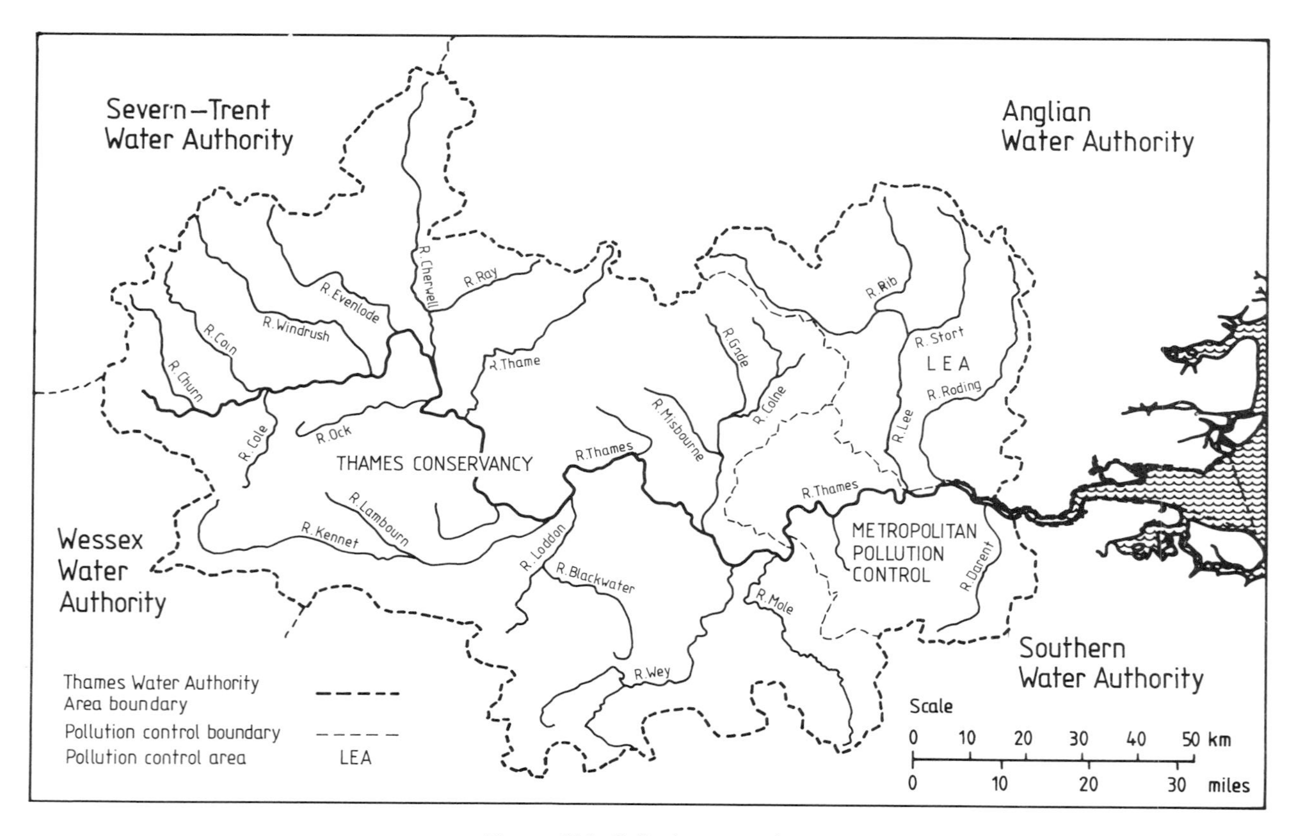

Figure 106. Pollution control map.

Table 37. Frequency of sampling determinands in the Thames Tideway and metropolitan watercourses†.

Determinand	Frequency of sampling			
	Tideway	Metropolitan watercourses	Effluents	
			Sewage works	Industrial
Temperature (water)	W	W	M	M
Temperature (air)	W	—	—	—
Suspended solids (total)	W	M	M	M
Suspended solids (non-volatile)	W	M	M	M
Chlorion (Cl^-)	W	M	M	—
Ammonia nitrogen	W	M	M	M
Ammonia (non-ionised)	W	W	—	—
Nitrite nitrogen	W	M	M	—
Nitrate nitrogen	W	M	M	—
BOD (ATU)	W	W	M	M
Dissolved oxygen	W	W	M	—
pH	W	W	M	M
Organic carbon	W	—	—	S
Chlorophyll '*a*'	W	—	—	—
Surfactant anionic	Q	Q	M	S
Surfactant non-ionic	Q	—	S	S
Hardness	—	Q	—	—
Conductivity	—	Q	—	—
Phosphate (soluble)	Q	Q	S	—
Chemical oxygen demand (COD)	—	Q	M	—
Mercury	Q	6M	M	S
Other metals (Zn, Cu, Ni, Pb, Cd)	Q	6M†	M	S
Sulphate	—	6M	—	—
Organohalogens	Q	S	S	S
Pesticides	Q	S	S	S
PCB's	Q	S	S	S
E. Coli	M	—	S	—
Phytoplankton	M	—	—	—
Organic nitrogen	—	M	M	—
Carbohydrate	—	—	—	S
Biological survey	M	M	—	—

† Chromium and iron in addition.

Frequency code:
W = not less than weekly
M = not less than monthly
6M = not less than 6-monthly
Q = not less than quarterly
S = sporadic.

Pollution Control

The Tideway receives large volumes of industrial discharges, but most of these now comprise unpolluted, recirculated cooling water. All are regulated to stringent quality standards, and all are monitored regularly. The monitoring of the discharges to the Tideway and of the water itself have been carried out regularly for a very long time (figure 106). Since the work was started by W J Dibdin in 1893, river water from Teddington to the spoil grounds in the Barrow Deep has been sampled not less than once a week. The ships (sludge vessels) which carry the cargoes of sludge are fitted with laboratories whereby those analyses requiring to be made immediately are carried out; the remainder are made on samples taken ashore.

Samples downstream of the main metropolitan sewage works are also taken in this way. Upstream samples from Beckton to Richmond are taken and analysed on special launches in a similar manner. All analytical results are presented at half-tide positions. Because the sludge vessels have fixed courses and cannot stop, their laboratories are unsuitable for some biological survey work so that special launches which can go safely into the outer estuary are used for this purpose.

The main sewage effluents discharged to the Tideway are also monitored regularly. The works staff are required to do this daily for operational reasons, and note is taken of the results for pollution control purposes. However, in order than an organisation such as a water authority, which carries out both 'poacher' and 'gamekeeper' functions, can be seen to be accountable to the general public, 'audit' samples which assess the conformity of effluents and industrial discharges to statutory requirements are taken only by a pollution control section (Nicolson and Wood 1974). The pollution control authority for the Tideway also takes frequent samples of industrial discharges. The frequency of sampling is shown in table 37.

Appendix VI

Glossary of Terms

Abstraction
The removal of water from any source, either permanently or temporarily, so that (*a*) it ceases to be part of the resources of that area, or (*b*) is transferred to another source within the area.

Activated carbon treatment
A process for removing traces of organic substances and gases from water, e.g. for amelioration of taste, odour or colour problems in water, by adsorption and absorption on or into activated carbon.

Activated sludge
Accumulated biological mass (floc) produced in the treatment of waste water by the growth of bacteria and other micro-organisms in the presence of dissolved oxygen.

Activated sludge treatment
A process for the biological treatment of waste water in which a mixture of waste water and activated sludge is agitated and aerated. The activated sludge is subsequently separated from the treated waste water by sedimentation, and is removed or returned to the process as needed.

Aerobic
Requiring (or not destroyed by) the presence of free oxygen.

Albuminoid ammonia
The ammonia released from organic nitrogen compounds which are decomposed fairly easily by oxidation.

Anaerobic
Requiring (or not destroyed by) the absence of free oxygen.

Aquifer
Porous water-bearing formation (bed or stratum) of permeable rock, chalk, sand or gravel capable of yielding significant quantities of water.

Benthic deposit
An accumulation on the bed of a watercourse or lake or the sea of deposits, possibly containing organic matter and arising from such causes as natural erosion, biological processes or discharge of waste water.

Benthos
Organisms living in or on the bottom of a water body.

Biodegradation
The molecular degradation of organic matter by the action of micro-organisms, leading to partial or complete mineralisation.

Biological, trickling or percolating filters
A bed of fragments of inert material through which waste water is caused to percolate for the purpose of purification by means of an active biological film on the inert material.

Biota
The living component of any aquatic system.

BOD (*biochemical oxygen demand*)
The mass concentration of dissolved oxygen consumed under specified conditions by the biological oxidation of organic matter in water.

BOD (ATU)
As for BOD but nitrification inhibited by addition of allyl thiourea to the sample.

COD (*chemical oxygen demand*)
The mass of oxygen equivalent to the amount of a specified oxidant consumed by dissolved or suspended matter when treating a water sample with the oxidant under defined conditions.

Detritus
In a biological context, organic particulate matter. In the context of sewage treatment practice, coarse debris denser than water but capable of being transported in moving water.

Denitrification
The liberation of nitrogen or nitrous oxide from nitrogenous compounds (particularly nitrates and nitrites) in water or waste water usually by the action of bacteria.

Deoxygenation
The partial or complete removal of dissolved oxygen from water, either under natural conditions or deliberately by physical or chemical processes.

Dewatering
The process whereby wet sludge, sometimes conditioned by a coagulant, has its water content reduced by physical means.

Digestion
The stabilisation, by biological processes of organic matter in sludge, normally by an anaerobic process.

Ecology
The study of the interrelation between living organisms and their environment.

Ecosystem
An ecological system in which, by the interaction between the different organisms present and their environment, there is a cyclic interchange of materials and energy.

Effluent polishing
Tertiary treatment employing either further physical or biological processes.

EOL *(effective oxygen load)*
The polluting load, expressed in terms of oxygen consumed, attributable to carbonaceous and nitrogenous matter. Calculated from: EOL = flow × [1.5 BOD + 4.5 (ammonia + organic nitrogen)].

Eutrophication
The enrichment of water, both fresh and saline, by chemical substances, especially compounds of nitrogen and phosphorus or nutrients that accelerate the growth of algae and higher forms of plant life.

Humus sludge
The microbial film which sloughs off from a biological filter and leaves with the trickling filter effluent and is normally separated in a final settling tank.

Macro-invertebrates
Larger (i.e. visible to the naked eye, not microscopic) animals without backbones.

Macrophytes
Large water plants other than algae, mosses and liverworts.

Mass balance
The relation between input and output of a specified substance in a defined system e.g. in a lake, river or sewage treatment works.

Methaemoglobinaemia
A cyanosis of babies which may occur within the first few weeks after birth, associated with the ingestion of excessive quantities of nitrate, e.g. by consuming nitrate-rich water.

Mixed liquor
A mixture of settled sewage and activated sludge undergoing circulation and aeration in the aeration tank or channel of an activated sludge plant.

Mixed liquor suspended solids (MLSS)
The concentration of solids, expressed in a specified dried form, in the mixed liquor.

Nitrification
The oxidation of nitrogenous matter by bacteria. Usually, the end products of such an oxidation are nitrates.

Oxygen saturation value, C_s
The concentration of dissolved oxygen in equilibrium with air (in natural systems) or pure oxygen (in oxygen waste water treatment systems); it varies with temperature, partial pressure of oxygen and salinity.

Permanganate value (PV)
The oxygen absorbed (mg/l) from acidified *N*/80 potassium permanganate in 4 hours at 27°C.

Photosynthesis
The synthesis of organic matter, by living organisms, employing photochemically reactive pigments, from carbon dioxide and water in the presence of light.

Ponding
The occurrence of pools of liquid on a biological filter caused by blockage of its interstices.

Preliminary treatment
The removal or disintegration of gross solids in sewage and the removal of grit, also sometimes the removal of oil and grease from sewage prior to sedimentation.

Primary treatment
The first major stage of treatment following preliminary treatment in a sewage works, usually involving the removal of settleable solids.

Retention period (detention time)
The theoretical period during which water or waste water is retained in a particular unit or system as calculated from a specified flow.

Run-off
The discharge of water derived from rain or snow falling on a surface.

Secondary treatment
The treatment of sewage, usually after the removal of suspended solids, by bacteria under aerobic conditions during which organic matter in solution is oxidised or incorporated into cells which may be removed by settlement. This may be achieved by biological filtration or by the activated sludge process.

Sedimentation
The process by which settleable solids are removed from waste water by passing it through a tank at such a velocity that the solids gravitate to the floor to form a sludge.

Seeding
The inoculation of a biological system for the process of introducing appropriate micro-organisms.

Self-purification
The process whereby polluting materials discharged to a natural water system are removed by physicochemical and biological agencies within the system itself.

Septic
A condition produced by putrefaction, resulting in the absence of dissolved oxygen.

Septic tank
Closed sedimentation tank in which settled sludge is in immediate contact with the waste water flowing through the tank, and the organic solids are decomposed by anaerobic bacterial action.

Settleable solids
That proportion of the initially suspended solids which will no longer remain in suspension after a specified settling period.

Sewage
The water-borne wastes of a community.

Sewerage
A system of pipes and appurtenances for the collection and transportation of domestic and industrial waste waters.

Sloughing
The continuous release of zoogloeal film material from the supporting media of a biological filter, in the form of humus sludge.

Slow sand filtration
A water-treatment process where large areas of sand are flooded with water and the physical, chemical and biological process of filtration results in purified percolate. This process is also sometimes used for the final treatment of sewage effluent after passing through conventional treatment facilities.

Sludge
A mixture of solids and water produced during the treatment of waste water.

Storm sewage
A mixture of sewage and the surface water arising from heavy rainfall or melting snow (ice).

Storm water: storm water run-off
Surface water draining to a watercourse as a result of heavy rainfall.

Supernatant liquor
The liquor in a sedimentation tank or settlement tank, or a sludge digestion tank, lying between the deposited solids and any floating scum.

Tertiary treatment
The application of additional treatment processes to reduce further the polluting effects of sewage which has undergone primary and secondary treatment. This may refer to (1) further physical treatment; (2) chemical treatment; or (3) further biological treatment.

Thickening
The process of increasing the concentration of solids in a sludge by the removal of water.

Treated sewage
Sewage that has received partial or complete treatment for the removal and mineralisation of organic and other material.

References

Andrews M J 1977 Observations on the fauna of the Metropolitan River Thames during the drought in 1976 *London Naturalist* **56** 46

Andrews M J and Rickard D G 1980 Rehabilitation of the Inner Thames Estuary *Marine Pollution Bulletin* **11** 327–32

Baker B and Binnie A R 1891 *Main Drainage of London* London County Council

Barclay W G 1963 Pollution of the Thames Valley during the last century *Effluent and Water Treatment Convention* (London: Seymour Hall)

Barnard J L 1973 Biological denitrification *Water Pollution Control* **6** 705

Barrett M J, Mollowney B M and Casapieri P 1978 The Thames model: an assessment *Progress in Water Technology* **10** 409

Barton N J 1962 *The Lost Rivers of London* (London: Phoenix House)

Berry G C 1957 Sir Hugh Myddleton and the New River *Transactions of the Honourable Society of Cymmrodorion, London*

Binnie A R 1899 *A Report upon the Main Drainage of London* London County Council

Chadwick E 1842 *The Poor Law Commission on an Inquiry into the Sanitary Conditions of the Labouring Population of Great Britain*

Clayton R 1964 *The Geography of Greater London* (London: George Philip and Sons)

Clowes F and Houston A C 1904 *The Experimental Bacterial Treatment of London Sewage* London County Council

Cooper P F, Drew E A, Bailey D A and Thomas E V 1977 Recent advances in sewage effluent denitrification: Part 1 *Water Pollution Control* **76** (3) 287

Cornish C J 1902 *The Naturalist on the Thames* (London: Seeley)

Dainty S H, Drake R A R, James J E and Shephard T E 1972 Design of extensions to the Beckton sewage treatment works of the Greater London Council *Proc. Inst. Civil Engineers* **51** 657

Dibdin W J 1887 *Report to the Metropolitan Board of Works* Appendix 2

—— 1894 *Results of Examination of the Character of the Water of the River Thames from Teddington to the Nore* London County Council

—— 1896 Purification of the Thames *Proc. Inst. Civil Engineers* **129**

Drake, R A R, Scott J, Clark T A and Cunningham J B 1977 Construction of extensions to Beckton sewage treatment works, Thames Water Authority *Proc. Inst. Civil Engineers* **62** 349

EEC 1973 *Directive on the method of control of the biodegradability of anionic surfactants* 22-11-72, OJ L347 (17-12 73)

Field R I, Moffa P E and MacArthur D A 1977 Treatability determinations for a prototype swirl combined sewer overflow regulator/solids separator *Progress in Water Technology* **8** (6)

Fitter R S R 1945 *London's Natural History* (London: Collins)

Fitzstephen W (*c* 1180) *Vita Sancti Thomea* (Preface)

Gameson A L H and Robertson K G 1955 *J. Applied Chemistry* 5 502

Gray J S and Mirza F B 1979 A possible method for the detection of pollution induced disturbances on marine benthic communities *Marine Pollution Bulletin* **10** 142

Greater London Council 1969, 1970 *Thames Flood Prevention* First report of studies October 1969; second report of studies 1969–70

Harington J 1596 *A New Discourse of a Stale Subject called the Metamorphosis of Ajax*

Horner R W 1978 The Thames tidal flood works in the London excluded area *Public Health Engineer* **6** (1) 16

Imhoff K R and Albrecht D 1977 Pure oxygen aeration in the Ruhr River *J. Water Pollution Control Fed.* September

Jonson B 1616 *On the Famous Voyage*

Kekwick A 1965 The Soho epidemic of cholera *Medico-legal Journal* 13 May p153

Kooijmans L P L 1972 Mesolithic bone and antler implements from the North Sea and from the Netherlands *Bericht van de Rijksdienst voor het Oudheidkundig Bademonderzoek* **20** (1) 27

Ministry of Housing and Local Government 1961 *Pollution of the Tidal Thames* (Pippard Report) (London: HMSO)

Monod J 1942 *Recherches sur la Croissance des Cultures Bactériennes* (Paris: Herman)

National Water Council 1976 *Report of Working Party on Consent Conditions for Effluent Discharges to Freshwater Streams* (London)

Nicolson N J and Wood L B 1974 The impartial assessment of performance of the sewage treatment works of a regional water authority *Municipal Engineering Supplement* October

Prus-Chacinski T M 1976 *Study of an Experimental Spiral Storm Overflow for Sewage, Technical Note* 77 (Construction Industry Research and Information Association)

Punch 1842 *The Thames and its Tributaries* **3** 165 (July–December)

Rawlinson R 1870 *Report upon Inquiry as to the Truth or Otherwise of Certain Allegations Contained in a Memorial from the Vicar and other Inhabitants of Barking in the County of Essex, Calling Attention to the Pollution of the River Thames by the Discharge of Sewage through the Northern Main Outfall Sewer of the Metropolitan Board of Works* (London: HMSO)

Regan C J 1951 Early developments in London's sewage disposal with particular reference to the work of W.J. Dibdin *J. and Proc. Inst. of Sewage Purification* 338

Royal Commission on Environmental Pollution 1972 *Third Report* (London: HMSO)

Sheldon D 1979 Private communication

Stow J 1598 *Survey of London*

Taylor J R, Humphreys Sir Geo. W and Peirson Frank T 1935 *Greater London Drainage* (London: HMSO)

Thames Water Authority 1977 *Thames Migratory Fish Committee Report*

—— 1978 *Report of Working Party on Thames Migratory Salmonids* Regional Fisheries Advisory Committee 7 April

Thurston G 1964 *The Great Thames Disaster* (London: Allen and Unwin)

Tittley I and Price J H 1977 The marine algae of the tidal Thames *London Naturalist* **56** 10

Townend C B 1962 Reflections on twenty-five years in Middlesex *J. and Proc. Inst. of Sewage Purification* **1** 19

Trevelyan G M 1948 *English Social History* (London: Longmans)

Vick E H 1967 The Crossness sewage treatment works of the Greater London Council *Water Pollution Control* **66** 82

Water Pollution Research Laboratory 1964 *Effects of Polluting Discharges on the Thames Estuary* (London: HMSO)

Watson J D 1929 West Middlesex sewerage and sewage disposal *Report to Middlesex County Council* 3 January

Welch C 1894 *History of the Tower Bridge* (London: Smith, Elder & Co)

Wheeler A 1979 *The Tidal Thames* (London: Routledge and Kegan Paul)

Williams A C 1946 *Angling Diversions* (London: Herbert Jenkins)

Wood L B 1977 River standards of the Thames Water Authority (2) Standards for discharge to the Thames tideway *Progress in Water Technology* **8** (6)

Wood L B, Borrows P F and Whiteland M R 1979a Scheme for remedying the effects of storm sewage overflows to the tidal River Thames *Workshop on Treatment of Domestic and Industrial Wastewater in Large Plants* (Vienna: Int. Association of Water Pollution Research), 1980 *Progress in Water Technology* **12** (3) 93–107

Wood L B and Cockburn A G 1979b An equitable approach to pollution control with particular reference to the Thames estuary *Workshop on Treatment of Domestic and Industrial Wastewater in Large Plants* (Vienna: Int. Association of Water Pollution Research), 1980 *Progress in Water Technology* **12** (3) 83–91

Wood L B, King R P, Durkin M K, Finch H J and Sheldon D 1976 The operation of a Simplex activated sludge pilot plant in an atmosphere of pure oxygen Part I *Public Health Engineer* **4** (2) 36; Part II *Public Health Engineer* **4** (3) 67

Wood L B and Morris H 1966 Modifications to the B.O.D. test *J. and Proc. Inst. of Sewage Purification* **4** 350

Wood L B and Richardson M L 1979 Catchment quality control *Workshop on Treatment of Domestic and Industrial Wastewater in Large Plants* (Vienna: Int. Association of Water Pollution Research), 1980 *Progress in Water Technology* **12** (3) 1–12

Index

All references are to page numbers. Where the information referred to is contained entirely in a figure or table the page number is in italic type.